AF847616

Springer Proceedings in Physics

Volume 318

Indexed by Scopus

The series Springer Proceedings in Physics, founded in 1984, is devoted to timely reports of state-of-the-art developments in physics and related sciences. Typically based on material presented at conferences, workshops and similar scientific meetings, volumes published in this series will constitute a comprehensive up to date source of reference on a field or subfield of relevance in contemporary physics. Proposals must include the following:

- Name, place and date of the scientific meeting
- A link to the committees (local organization, international advisors etc.)
- Scientific description of the meeting
- List of invited/plenary speakers
- An estimate of the planned proceedings book parameters (number of pages/articles, requested number of bulk copies, submission deadline).

Please contact:

For Americas and Europe: Dr. Zachary Evenson; zachary.evenson@springer.com
For Asia, Australia and New Zealand: Dr. Loyola DSilva; loyola.dsilva@springer.com

Arthur Gibadullin · Geetha Devi · Shahriyor Sadullozoda
Editors

Proceedings of the 4th International Scientific and Practical Symposium on Materials Science and Physics

MST2024, December 23–25, 2024, Dushanbe, Tajikistan

Springer

Editors
Arthur Gibadullin
National Research University
Moscow Power Engineering Institute
Moscow, Russia

Geetha Devi
National University of Science
and Technology
Muscat, Oman

Shahriyor Sadullozoda
Khujand Polytechnic Institute of Tajik
Technical University Named After
Academician M.S.Osimi
Khujand, Tajikistan

ISSN 0930-8989 ISSN 1867-4941 (electronic)
Springer Proceedings in Physics
ISBN 978-3-031-94804-6 ISBN 978-3-031-94805-3 (eBook)
https://doi.org/10.1007/978-3-031-94805-3

© The Editor(s) (if applicable) and The Author(s), under exclusive license to Springer Nature Switzerland AG 2025

This work is subject to copyright. All rights are solely and exclusively licensed by the Publisher, whether the whole or part of the material is concerned, specifically the rights of translation, reprinting, reuse of illustrations, recitation, broadcasting, reproduction on microfilms or in any other physical way, and transmission or information storage and retrieval, electronic adaptation, computer software, or by similar or dissimilar methodology now known or hereafter developed.
The use of general descriptive names, registered names, trademarks, service marks, etc. in this publication does not imply, even in the absence of a specific statement, that such names are exempt from the relevant protective laws and regulations and therefore free for general use.
The publisher, the authors and the editors are safe to assume that the advice and information in this book are believed to be true and accurate at the date of publication. Neither the publisher nor the authors or the editors give a warranty, expressed or implied, with respect to the material contained herein or for any errors or omissions that may have been made. The publisher remains neutral with regard to jurisdictional claims in published maps and institutional affiliations.

This Springer imprint is published by the registered company Springer Nature Switzerland AG
The registered company address is: Gewerbestrasse 11, 6330 Cham, Switzerland

If disposing of this product, please recycle the paper.

Organization

Program Committee Chairs

Geetha Devi, National University of Science and Technology, Oman
Shahriyor Sadullozoda, Khujand Polytechnic Institute of Tajik Technical University named after academician M.S. Osimi, Tajikistan
Arthur Gibadullin, National Research University "Moscow Power Engineering Institute", Russia

Program Committee

Garlapati Chandrasekhar, Puducherry Technological University, India
Gulom Uzakov, Karshi State Technical University, Uzbekistan
Dmitry Morkovkin, Financial University under the Government of the Russian Federation, Russia
Dilnoz Muhamediyeva, Tashkent Institute of Irrigation and Agricultural Mechanization Engineers National Research University, Uzbekistan
Yerbol Ospanov, Shakarim University, Kazakhstan
Sergey Voinash, Kazan Federal University, Russia
Valentina Britvina, Moscow Polytechnic University, Russia
Lola Safarova, Samarkand State University of Veterinary Medicine, Livestock and Biotechnologies, Uzbekistan
Sayyora Khamraeva, Karshi State Technical University, Uzbekistan
Almaz Mehdiyeva, Azerbaijan State University of Oil and Industry, Baku, Azerbaijan
Manuchehr Sadriddinov, International University of Tourism and Entrepreneurship of Tajikistan, Tajikistan
Narzullo Mamatov, Tashkent Institute of Irrigation and Agricultural Mechanization Engineers National Research University, Uzbekistan
Zariya Zalilova, Bashkir State Agrarian University, Russia

Valentina Pulyaeva, Financial University under the Government of the Russian Federation, Moscow, Russia
Ramazon Abdullozoda, Tajik Technical University named after academician M.S. Osimi, Tajikistan

Organizer

Tajik Technical University named after Academician M.S. Osimi, Dushanbe, Tajikistan

Preface

The organizing team is proud to publish the selected papers of the IV International Symposium, (MST2024), held from 23 to 25 December 2024 in the Springer Proceedings in Physics Series. The MST2024 Symposium intends to be an international platform for the potential researchers and materials engineers to present and showcase the recent developments and techniques in the field of physics, materials science, biological, chemical and mechanical materials. All the submitted papers were undergone three extensive peer reviews by qualified experts as independent reviewers in the field selected by the symposium committee.

As a result of the work of all reviewers, out of 32 received papers, the top 12 papers were accepted for the publication. The peer review focused on the quality of the content, the relevance of the research, novelty and scientific significance and the originality of the work. Reviewers, members of the program committee and members of the organizing committee were kept away from the peer review process to avoid any conflicts of interest.

The symposium was intended to provide international experience of the participants in the field of physics and materials science. In particular, the following issues were considered:

- Formation of scientific and practical proposals for the development of materials science, chemistry, new polymers and composites;
- Development of fundamental and applied research in the field of chemical physics and physical chemistry and related sciences;
- Formation of recommendations aimed at improving the quality of materials and developing the scientific and technical base of new materials and composites.

In the symposium, the authors of the articles presented and displayed their novel research findings for broad discussion, searching for options for interaction and gather new approaches for the further development of their research.

A distinctive feature of the symposium is that authors and speakers were from the different nationals like People's Republic of China, the Sultanate of Oman, the Republic of Azerbaijan, the Russian Federation and the Republic of Uzbekistan spoke at it. Researchers from different countries presented their research outcome and achievements in the field of physics, materials science and new materials.

Thus, the symposium offered knowledge exchange and gain advanced knowledge for all participants, and the results of the symposium were beneficial for state and regional authorities for the development of industry and new areas of activity, international and supranational organizations, scientific and professional community.

The organizing committee of the symposium expresses their gratitude to the Springer staff that supported the publication of this collection. In addition, the organizing committee would like to thank the symposium participants, reviewers and those who helped to successfully organize this symposium.

Moscow, Russia — Arthur Gibadullin
Muscat, Oman — Geetha Devi
Khujand, Tajikistan — Shahriyor Sadullozoda

Contents

1 **Changes in the Spectrum of Acoustic Emission Signal in the Material Fiberglass Under the Influence of Tensile Stresses** 1
O. V. Bashkov, I. O. Bashkov, M. N. Bezruk, T. A. Efimov, R. V. Romashko, and F. Xiao

2 **Physical Method for Determining the Parameters of a Hydraulically Satisfactory Reliable Cross-Section of Main Channels** 13
Ahliman Verdiyev and Maya Abdullayeva

3 **Physical Principles of Using New Promising Materials and Technologies for Environmentally Friendly and Safe Nuclear Power Reactors** 25
V. S. Okunev

4 **Impact Toughness and Tensile Strength Tests of $AlSi_{10}Mg$ Alloy Specimens After Selective Laser Melting** 57
Artyom S. Raznoschikov, Leonid V. Belyaev, Alexey V. Zhdanov, and Nikita S. Dovbysh

5 **Physicochemical Preparation for Fabrication and Characterization of Gold Nanoparticles for Antibacterial Activity Study** 69
Saif Al Busaidi, Geetha Devi, and Qais Mohammed Mahmood Al Balushi

6 **Application of the Metaheuristic Algorithms of Global Optimization for Determination of the Composite Materials Constituent's Properties** 79
Andrei Panteleev, Nikolay Turbin, and Maksim Mukhai

7 Physical Principles of Wastewater Heat Utilization Using High-Temperature Solar Installations 95
Gulom Uzakov, Bobir Toshmamatov, and Nasiba Abdurakhmanova

8 Local Sodium Carboxymethyl Cellulose: Physicochemical Characteristics and Their Significance 111
Ummatjon A. Asrorov, Sabitjan Y. Inagamov, Erkin B. Khujanov, Furqat K. Turotov, and Hasan N. Bozorov

9 A New Approach to the Analysis of Nuclear Reactions Under the Influence of High-Energy Particles on a Stationary Target Nucleus in the Physics of Resonant Nuclear Reactions 125
Ibratjon Aliyev, G'olibjon Qo'ldashov, Sultonali Abdurakhmonov, Inomjon Bilolov, and Sharobiddin Isroilov

10 Some Questions of Thermonuclear Fusion Physics: The New Mechanism of the ^{2}H-^{3}H Reaction 143
V. S. Okunev

11 Research of the Physical Basis of Heat Transfer in Heat Exchanger Devices 167
Saydulla Khuzhakulov, Zakir Pardaev, Ismoildjon Muradov, and Timur Shodiev

12 Mathematical Modeling of the Phenomenon of Electrical Conductivity of a Lithium-Ion Battery with a Large Capacity 183
Ibratjon Aliyev, Sultonali Abdurakhmonov, Ruzikhon Akhrorova, Moxizarxon Mamadaliyeva, and Nargiza Toirova

List of Contributors

Maya Abdullayeva Azerbaijan State Oil and Industry University, Baku, Azerbaijan

Nasiba Abdurakhmanova "Shurtan GChC" LLC, Guzar d, Uzbekistan

Sultonali Abdurakhmonov Fergana Polytechnic Institute, Fergana, Fergana Region, Republic of Uzbekistan;
Ferghana State Technical University, Fergana, Fergana Region, Republic of Uzbekistan

Ruzikhon Akhrorova Andijan State University of Foreign Languages, Andijan, Fergana Region, Republic of Uzbekistan

Qais Mohammed Mahmood Al Balushi College of Engineering, National University of Science and Technology, Muscat, Sultanate of Oman

Saif Al Busaidi College of Engineering, National University of Science and Technology, Muscat, Sultanate of Oman

Ibratjon Aliyev SRI "PRNR", Electron Laboratory LLC, Margilan, Fergana Region, Republic of Uzbekistan

Ummatjon A. Asrorov National University of Uzbekistan Named After Mirzo Ulugbek, Tashkent, Uzbekistan;
Tashkent State Pedagogical University Named After Nizami, Tashkent, Uzbekistan

I. O. Bashkov Komsomolsk-na-Amure State University, Komsomolsk-on-Amur, Russia

O. V. Bashkov Komsomolsk-na-Amure State University, Komsomolsk-on-Amur, Russia

Leonid V. Belyaev Vladimir State University named after Alexander and Nikolay Stoletovs, Vladimir, Russia

M. N. Bezruk Institute of Automation and Control Processes, FEB RAS, Vladivostok, Russia

Inomjon Bilolov Ferghana State Technical University, Fergana, Fergana Region, Republic of Uzbekistan

Hasan N. Bozorov Tashkent State Pedagogical University Named After Nizami, Tashkent, Uzbekistan

Geetha Devi College of Engineering, National University of Science and Technology, Muscat, Sultanate of Oman

Nikita S. Dovbysh Vladimir State University named after Alexander and Nikolay Stoletovs, Vladimir, Russia

T. A. Efimov Institute of Automation and Control Processes, FEB RAS, Vladivostok, Russia

Sabitjan Y. Inagamov Tashkent Pharmaceutical Institute, Tashkent, Uzbekistan

Sharobiddin Isroilov Ferghana State Technical University, Fergana, Fergana Region, Republic of Uzbekistan

Erkin B. Khujanov Tashkent State Pedagogical University Named After Nizami, Tashkent, Uzbekistan

Saydulla Khuzhakulov Karshi State Technical University, Karshi, Uzbekistan

Moxizarxon Mamadaliyeva Fergana State University, Fergana, Fergana Region, Republic of Uzbekistan

Maksim Mukhai Moscow Aviation Institute (National Research University), Moscow, Russia

Ismoildjon Muradov Karshi State Technical University, Karshi, Uzbekistan

V. S. Okunev Bauman Moscow State Technical University, Moscow, Russia

Andrei Panteleev Moscow Aviation Institute (National Research University), Moscow, Russia

Zakir Pardaev Karshi State Technical University, Karshi, Uzbekistan

G'olibjon Qo'ldashov National Research of "Renewable Energy Sources" Under the Ministry of Energy of the Republic of Uzbekistan, Tashkent Region, Tashkent, Republic of Uzbekistan

Artyom S. Raznoschikov Vladimir State University named after Alexander and Nikolay Stoletovs, Vladimir, Russia

R. V. Romashko Institute of Automation and Control Processes, FEB RAS, Vladivostok, Russia

Timur Shodiev Karshi State Technical University, Karshi, Uzbekistan

Nargiza Toirova Fergana State University, Fergana, Fergana Region, Republic of Uzbekistan

Bobir Toshmamatov Karshi State Technical University, Karshi, Uzbekistan

Nikolay Turbin Moscow Aviation Institute (National Research University), Moscow, Russia

Furqat K. Turotov National University of Uzbekistan Named After Mirzo Ulugbek, Tashkent, Uzbekistan

Gulom Uzakov Karshi State Technical University, Karshi, Uzbekistan

Ahliman Verdiyev "Water and Amelioration Scientific Research Institute", Public Legal Entity, Baku, Azerbaijan

F. Xiao Heilongjiang University of Science and Technology, Songbei District, Harbin, China

Alexey V. Zhdanov Vladimir State University named after Alexander and Nikolay Stoletovs, Vladimir, Russia

Chapter 1
Changes in the Spectrum of Acoustic Emission Signal in the Material Fiberglass Under the Influence of Tensile Stresses

O. V. Bashkov, I. O. Bashkov, M. N. Bezruk, T. A. Efimov, R. V. Romashko, and F. Xiao

Abstract The aim of the work is to establish the effect of tensile stresses in fiberglass on the spectra of acoustic emission signals recorded by a piezoelectric and fiber-optic acoustic emission sensor. For this purpose, fiberglass samples were made by molding using the vacuum infusion method. The samples were cut out for testing on a tensile testing machine. A fiber-optic sensor operating in conjunction with an adaptive laser holographic interferometer was glued to the surface of the working part of the samples. The piezoelectric transducer was installed on the non-working part of the sample. Acoustic emission signals were recorded by both sensors under stepwise loading with a tensile load. The result of the work is the analysis of changes in the spectra of acoustic emission signals caused by tensile stresses in the fiberglass material. It is noted that a shift in the spectrum to the high-frequency region with increasing stress is observed at low-frequency harmonics. At high frequencies that are informative for acoustic emission, the influence of tensile stresses becomes nonlinear.

1.1 Introduction

Innovative sensors based on optical methods are increasingly being used in industry. For example, fiber-optic sensors (FOS) based on Bragg diffraction gratings have been developed in the field of temperature and strain measurement [1–3]. The main advantage of fiber-optic sensors is their high protection from electromagnetic noise,

O. V. Bashkov · I. O. Bashkov
Komsomolsk-na-Amure State University, Komsomolsk-on-Amur, Russia

M. N. Bezruk · T. A. Efimov · R. V. Romashko
Institute of Automation and Control Processes, FEB RAS, Vladivostok, Russia

F. Xiao (✉)
Heilongjiang University of Science and Technology, Songbei District, Harbin, China
e-mail: xiaofukun1@usth.edu.cn

© The Author(s), under exclusive license to Springer Nature Switzerland AG 2025
A. Gibadullin et al. (eds.), *Proceedings of the 4th International Scientific and Practical Symposium on Materials Science and Physics*, Springer Proceedings in Physics 318,
https://doi.org/10.1007/978-3-031-94805-3_1

due to the lack of electronic signal transmission systems to the registration device [4, 5]. In addition, optical fibers can be embedded in the structure of the object's material during its manufacture or glued to the surface [6–8]. Certain facilities require constant monitoring of the technical condition in order to detect damage in a timely manner [9–12]. The acoustic emission (AE) method has these capabilities. The possibility of using optical fibers as AE wave sensors has previously been shown in the scientific literature [13–17]. Fiber-optic sensors on laser interferometers have higher sensitivity compared to sensors on Bragg arrays. This is justified by the distribution of the sensing element along the length of the entire fiber embedded in the object of control, when sensors on Bragg diffraction gratings are locally concentrated in certain places of the optical fiber. To combat temperature fluctuations and low-frequency mechanical noise, various methods of stabilizing the operating point are used in interference measuring systems.

One of the most effective methods is the method of adaptive laser holographic interferometry. The method is based on the interaction of the reference and object rays of the interferometer occurring in a photorefractive crystal [18–21]. As a result of the two-beam interaction, the hologram formed in the body of the photorefractive crystal is constantly overwritten. The rewriting time is a constant of the crystal, depends on its type and determines the frequency of suppression of low-frequency temperature and mechanical noise affecting the optical fiber measuring system. Thus, the adaptive interferometer generally works as a high-pass filter. This allows it to be used as a measuring system for recording AE signals. The frequency spectrum of acoustic waves generated by AE sources is quite wide. The spectrum of the recorded signal depends on the type of source, the propagation channel of the AE wave, the amplitude-frequency characteristics of electronic devices, amplifiers and converters. To eliminate acoustic noise in the sound range, the spectrum of analyzed AE signals is usually in the range above 20 kHz. Spectral methods of analysis of recorded AE signals are used to identify various types of damage in the controlled object. However, the spectral characteristics of signals in loaded structures can change, just as the frequency of vibrations of a stretched string varies depending on the degree of tension.

The work is devoted to the analysis of changes in the spectrum of the AE signal detected in loaded fiberglass by fiber-optic and piezoelectric converters.

1.2 Materials and Methods

1.2.1 Materials, Methods of Sample Manufacturing and Test Preparation

Only two levels of headings should be numbered. For the manufacture of fiberglass samples that could be used for tensile loading, it is necessary to produce blanks in the form of plates. Fiber-glass and binder are selected based on the requirements for the material of the fiberglass product. The work did not set the task of manufacturing samples with specified mechanical properties. Therefore, Umatex 1250-T30-290 fiberglass in the amount of 10 layers was used as a reinforcing material. The binder Derakane 411-350 was used as a polymer matrix. The plates were made by vacuum infusion with curing at room temperature.

After curing the workpiece material, samples with a rectangular cross section of the working area of 5 × 10 mm and a length of 100 mm were cut from the plate. The gripping points of the test machine were made in the form of a widening with a radius part on both sides. The optical fiber of the sensor of the adaptive interferometer measuring system was glued to the working surface of the sample with epoxy glue. In addition, a GT301 broad-band piezoelectric transducer (PZT) with a frequency band of 50–550 kHz was installed on the sample, connected to a broadband amplifier with a frequency band of 20–1000 kHz. The signals from FOS and PZT were recorded on a digital recording oscilloscope. The signals were recorded in the oscilloscope memory after exceeding the threshold for recording signals from PZT.

1.2.2 Methods of Testing and Data Analysis

The sample was subjected to stepwise tensile loading in a testing machine (see Fig. 1.1). When loaded to each next level, the load increased by 2500 N. Thus, the sample was loaded to 2500, 5000, 7500 N and so on until the sample was destroyed. At each stage, the loading was stopped in order to perform simulated excitation of the AE source on the sample. To do this, using a Hsu–Nielsen simulator, a pencil lead with a diameter of 0.5 mm and a hardness of 2H was broken on the surface of the sample. The waves generated by the AE simulator were recorded simultaneously by both sensors of the measuring systems and stored digitally on a recording digital oscilloscope.

After testing the sample until destruction, digital processing of the signals recorded by the FOS and PZT was performed. Digital processing included preliminary signal cleaning from high-frequency noise by applying filters. For noise removal and digital processing, the MatLab program tools and functions were used. The "*wdenoise*" function was used to remove noise from signals. By default, the "sym4" wavelet is used in the function, which is an orthogonal wavelet with a compact carrier with high symmetry. The "*UniversalThreshold*" noise removal method was selected.

At each of the loading stages, five AE signals recorded by FOS and PZT were saved. From the signals recorded at each loading stage by FOS and PZT, those with the maximum integral correlation coefficient among the others were selected. For further analysis, a discrete wavelet transform of signals by the orthogonal Daubechies wavelet of order 14 "db14" and the construction of the Fourier spectrum of signals and series of detailing coefficients of the discrete wavelet transform were used. Each of the selected signals was subjected to wavelet transformation by the Daubechies wavelet of order 14 at 10 decomposition levels. The last 10th level of decomposition

Fig. 1.1 A fiberglass sample installed in the test room at one of the loading stages

corresponds to a frequency band of 1–2 kHz at a sampling rate of 2 MHz of the recorded signals.

1.3 Results

1.3.1 Signals Denoising

The "*wdenoise*" function had a significant positive effect on the suppression of high-frequency noise of signals registered by FOS. Suppression of laser noise in the frequency range of 200–400 kHz smoothed the shape of the signals without changing their phase; amplitude and composition of low-frequency harmonics (see Fig. 1.2).

It should be noted that the informative frequency range of the AE signals visually detected by the harmonic spectrum recorded by FOS does not exceed 25 kHz. Denoising of the signals registered by the PZT did not significantly affect the waveform due to the fact that the signal from the PZT was sent to an amplifier with an analog filter installed in it, effectively suppressing high-frequency noises > 1000 kHz and low-frequency noises up to 20 kHz (see Fig. 1.3).

The spectra of AE signals recorded by FOS and PZT differ significantly, which is due to the characteristics of each converter. The PZT is a local sensor with its own amplitude-frequency characteristic (AFC), which depends on the material and

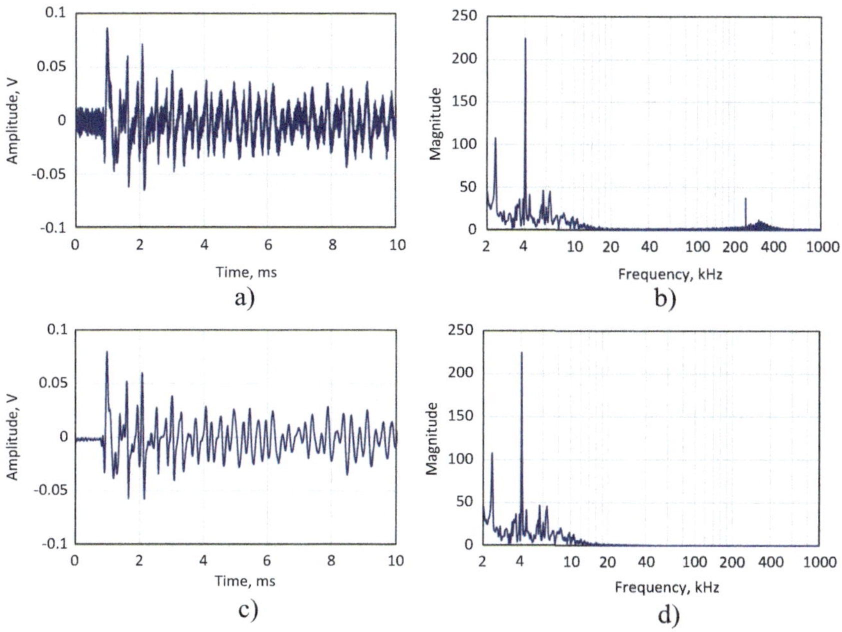

Fig. 1.2 AE signal and its spectrum registered by FOS before (**a**, **b**) and after (**c**, **d**) denoising using the "*wdenoise*" function

geometry of the piezoelectric plate located in the sensor body. PZT, installed on the sample, changes the spectrum of the signal with its AFC, which has undergone a transformation during the propagation of the acoustic wave from the source to the receiver. In addition, the signal from the PZT is amplified by an amplifier with its own AFC, which changes the spectrum of the recorded signal.

The frequency response of the interference system for measuring oscillations recorded by FOS is determined by two main elements. The first element is a photodetector that records laser radiation resulting from two-beam interaction occurring in a photorefractive crystal.

The interaction of the reference and information beams inside the photorefractive crystal results in a change in the radiation intensity level on the photodetector. The frequency response of the photodetector is determined mainly by the frequency response of the photodiode that perceives the radiation and is a fairly smooth characteristic with a decrease in the coefficient of conversion of the radiation intensity into electrical voltage with an increase in the oscillation frequency. The second element is the geometry of the optical fiber laying on the test object, combined with the geometry of the test object itself. The optical fiber placed on the sample perceives acoustic oscillations along the entire length of the fiber during the wave attenuation period simultaneously. The optical fiber experiences a superposition of stress and strain fields that form an additive change in the speckle at the output of the light

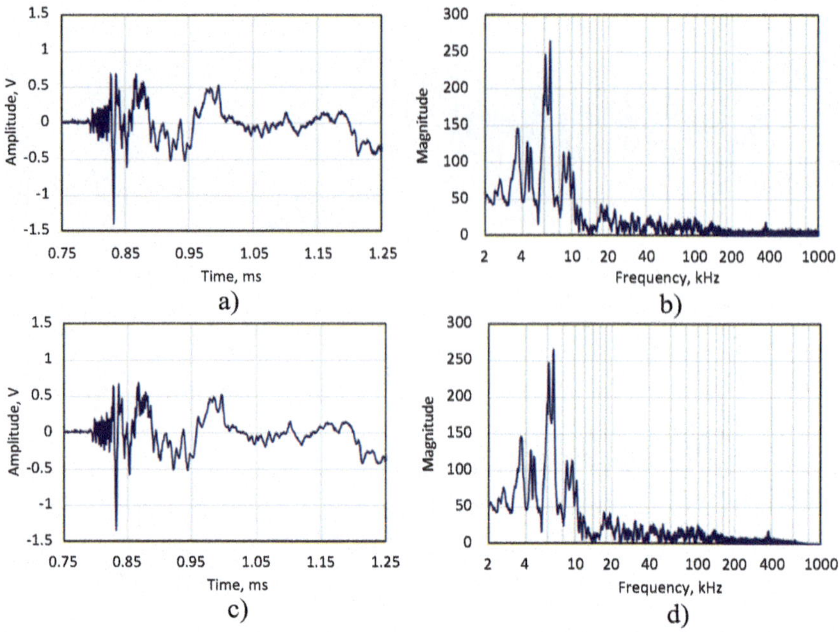

Fig. 1.3 AE signal and its spectrum registered by PZT before (**a**, **b**) and after (**c**, **d**) denoising using the "*wdenoise*" function

guide when the optical density changes in its individual parts. The incommensurate difference in the speeds of light radiation and sound waves leads to the fact that the phase change of the multimode light flow passing through the optical fiber occurs almost simultaneously with local deformation of its individual parts. Thus, the frequency response FOS is determined by the characteristics of the interference measuring system and the oscillatory properties of the object in which its sensitive element, perceiving oscillations—the optical fiber—is built in.

1.3.2 Signals Analyzing

Despite the difference in the spectra of the signals recorded by FOS and PZT, one can note the low-frequency nature of the signals as a whole, determined by the geometry of the sample. High values of the magnitudes of the signal spectra recorded by both sensors are in the frequency range of < 25 kHz, whereas the frequency range of the AE signals recorded in practice during the destruction of fiberglass is in the frequency range of up to 500 kHz. The sample is a waveguide in the form of a rod or a rectangular plate of limited size 5×10 mm. In waveguides of the rod and plate type, Pochhammer and Lamb waves are formed, mainly of low orders. The comparison of the signal spectra was performed separately for each sensor at different stages of loading.

A nonlinear change in the spectrum of the signals recorded by FOS is observed as the load on the sample increases. The magnitude of the low-frequency harmonic at a frequency of 1.8 kHz increases to 45–400 units with the load increasing to 12,500 N (see Fig. 1.4). The magnitude of the harmonic at a frequency of 4 kHz increases to a maximum value of 600 units as the load increases to 7500 N, then gradually decreases to 240 units. Similar changes are observed for the signals recorded by PZT in the spectral range of up to 3 kHz: the maximum magnitude in this range increases from 70 to 170 units as the load increases to 12,500 N (see Fig. 1.5). A non-monotonic change in magnitude is observed for the peaks of signal harmonics in the range of up to 10 kHz. At the same time, the magnitude level in the frequency range of 30–35 kHz smoothly increases from 20 to 60 units.

Due to the complexity of the spectral composition of the signals, the spectrum of the recorded AE signals was analyzed using 10 levels of wavelet decomposition. The 14th order Daubechies wavelet was used for decomposition. The frequency range above 1 MHz can be considered noise, so the first level of decomposition was excluded from the analysis. This approach to the analysis of the frequency components of the signal is justified primarily by the fact that when identifying the type of source, it is important to determine the type of destruction corresponding to the recorded AE signal.

When destroying polymer composites such as fiberglass and carbon fiber reinforced plastics, the main types of sources are the destruction of the polymer matrix of the binder and reinforcing glass or carbon fibers, as well as delamination of the layered composite material.

The spectral composition of the signals differs significantly, which allows them to be separated by the peaks of the frequency components located in different octaves. Based on the results of calculating the frequency components of the wavelet transform, graphs of the dependence of the reduced fraction of the root mean square (RMS) of the frequency component of the AE signal on the frequency were constructed. RMS is given as a percentage. Since the dynamic range of RMS values is quite large, the diagrams are shown in double logarithmic coordinates (see Fig. 1.6).

Considering that the RMS values are located quite close to each other, the diagrams are shown only for the loading stages of 0, 5000, 10,000 and 12,500 N. The sample failed upon reaching a load of 15,000 N. From the presented diagrams it can be seen that the most noticeable changes are observed upon transition from a load of 5000 to 10,000 N, for the signals recorded by both FOS and PZT. As was noted in the Fourier spectra (see Fig. 1.3), for the signals recorded by FOS, an increase in the RMS value is observed only for the range of 1–2 kHz upon changing the load of 12,500 N (see Fig. 1.5). The contribution of low-frequency components of 4–8 kHz to the energy of signals recorded by sensors of both types gradually decreases as the load of 10,000 N increases. In the ranges of 16–31 and 31–63 kHz, with an increase in load, the contribution of the specified frequency components to the signal energy increases by 1.48 times.

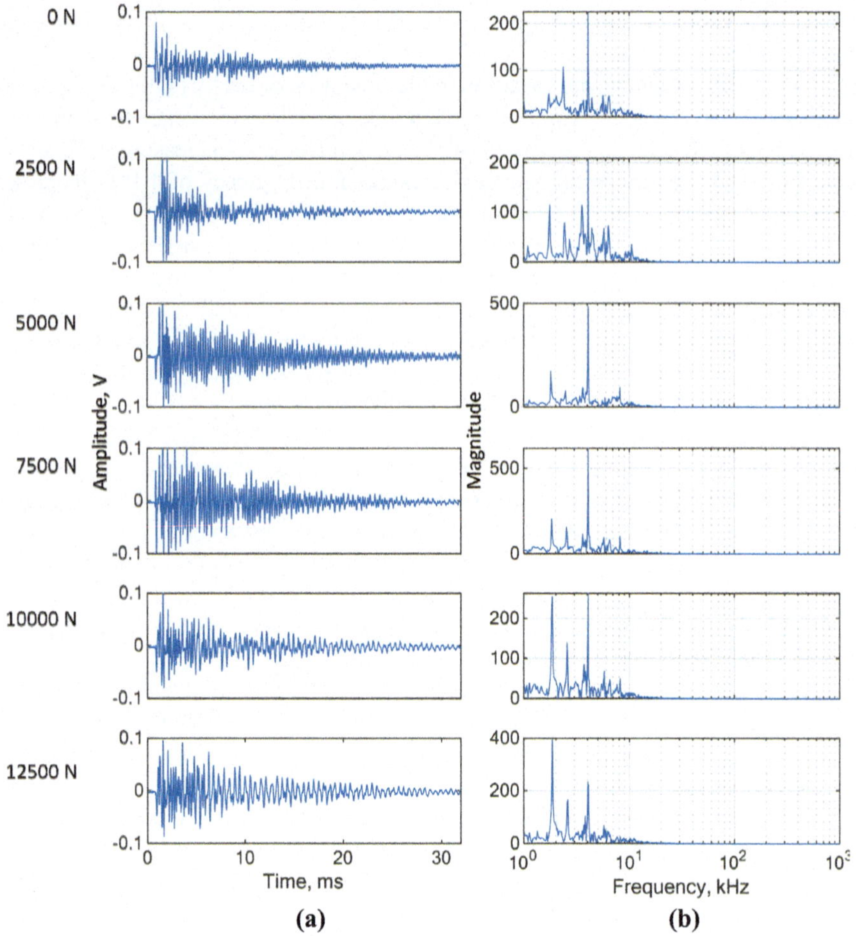

Fig. 1.4 AE signals **(a)** and their spectra **(b)** recorded by FOS at various loading stages

This frequency range for the composite material is typical for signal generation during destruction of the polymer matrix of the initial degree of damage. Although the radiation source in this case is not the matrix, but the fracture of the pencil, the change in the spectrum indicates that some local damage has already occurred in the composite material under this load, changing the spectrum of the transmitted sound wave.

Based on the characteristic grouping of the signal spectra recorded by both types of sensors under loads of 0 and 5000 N, as well as under loads of 10,000 and 12,500 N, one can note the similarity of the change in the spectral characteristics up to a frequency of 16 kHz (see Figs. 1.6 and 1.7). When the load exceeds 5000 N, the contribution of the 1–2 kHz component increases and the 4–8 and 8–16 kHz components decrease in the energy of the signals of both types of sensors.

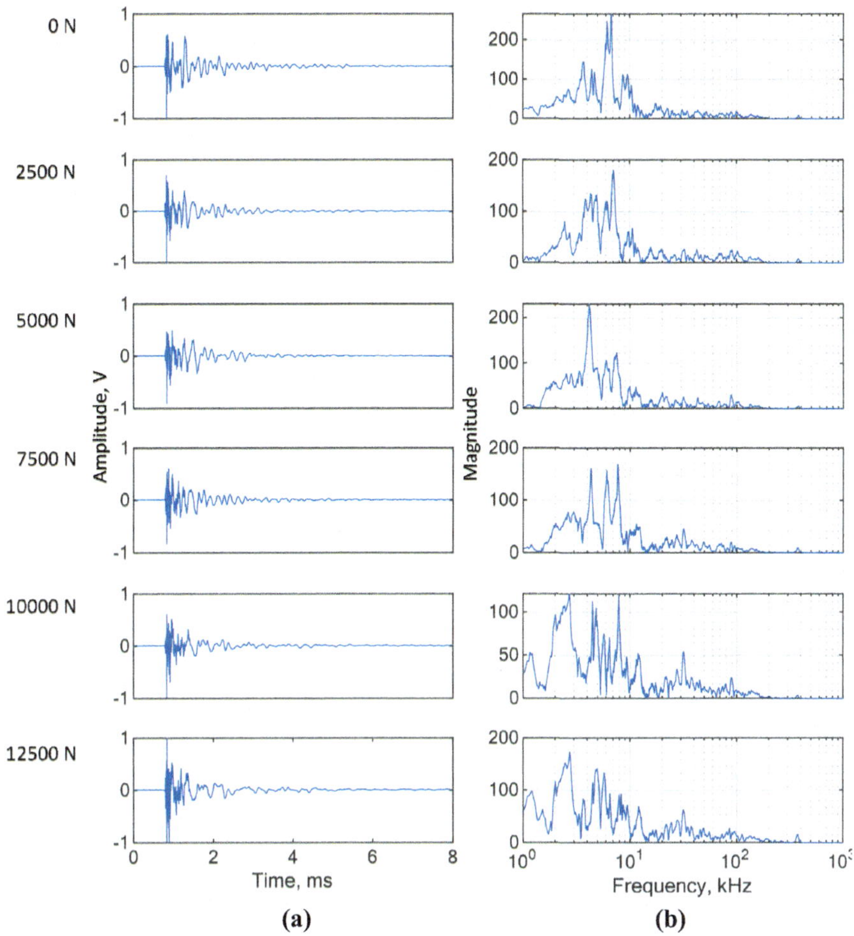

Fig. 1.5 AE signals (**a**) and their spectra (**b**) recorded by PZT at various loading stages

Considering that PZT is a local sensor, and FOS is distributed, it is logical to assume that the change in the signal spectrum is associated with an increase in the load in the sample, and not with a defect resulting from loading the sample. The spectrum change for signals recorded by FOS is less noticeable on the diagram due to the fact that the diagram is presented in a wider range of values due to high attenuation in the frequency range above 30 kHz. On the contrary, at each of the frequencies in the 8–31 kHz range, there is a gradual decrease in the RMS signal values as the load increases. This can be most clearly observed on the diagrams of the change in RMS values from the load (see Fig. 1.7).

It should be noted that the nature of the spectrum change for the frequency ranges 4–8 and 16–31 kHz has a qualitative similarity, although the contribution of higher frequency components to the signal composition is insignificant. When the load

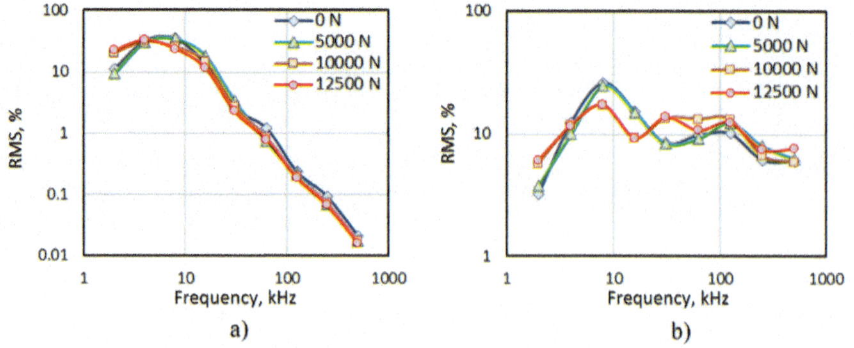

Fig. 1.6 The dependence of the reduced fraction of the RMS value (RMS) of the frequency component of the AE signal on the frequency: **(a)** FOS, **(b)** PZT

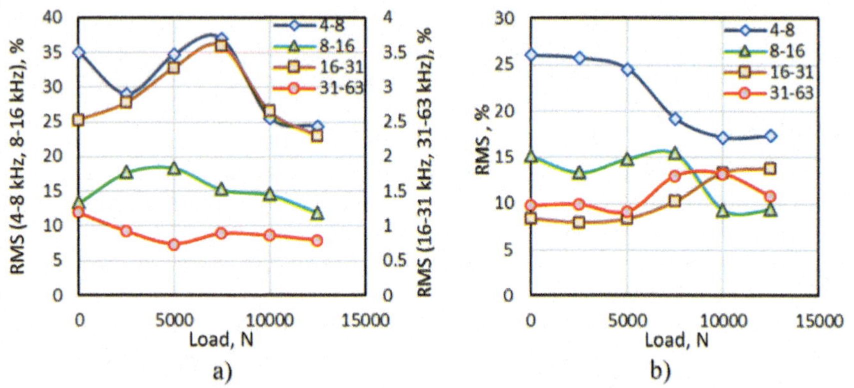

Fig. 1.7 The RMS dependence of the frequency component of the AE signal on the tensile load: **(a)** FOS, **(b)** PZT

increases from 2500 to 7500 N, RMS values increase by 1.3 times in these frequency ranges. A further step of increasing the load to 10,000 N leads to a decrease in RMS values by 1.4 times.

1.4 Conclusion

Thus, a change in the load in the composite material may be accompanied by structural damage while still in a state of elasticity. This leads to a change in the spectrum of the acoustic signal propagating in the control object. The difference in the types of transducers of PEP and WATER will not allow us to give a unified approach to describing the nature of changes in the spectrum of recorded AE signals. At the same time, a change in the signal spectrum may be the cause of damage accumulation in the material during its loading, and not caused by a change in the type of sources generated, which must be taken into account when analyzing the nature of damage to loaded objects.

Acknowledgements The work was supported by a grant from the Russian Science Foundation (Project No. 24-29-00838).

References

1. G. Kordas, G. Kordas, L.B. Liokumovich, N.A. Ushakov, Structural health monitoring with integrated optical fiber sensors: a review. AlfaBuild **4**(29), 2909 (2023)
2. W.W. Morey, G. Meltz, W.H. Glenn, Fiber optic Bragg grating sensors, in *Conference 1990, Fiber Optic and Laser Sensors VII, Proceedings SPIE*, vol. 1169 (1990)
3. D. Tosi, Review of chirped fiber Bragg grating (CFBG) fiber-optic sensors and their applications. Sensors **18**, 2147 (2018)
4. M. Muttillo, V. Stornelli, R. Paolucci, Structural health monitoring: an IoT sensor system for structural damage indicator evaluation. Sensors **20**(17), 1–15 (2020)
5. R. Hui, M. O'Sullivan, *Fiber-Optic Measurement Techniques* (Academic Press, 2022)
6. B. Culshaw, Fiber optics in sensing and measurement. IEEE J. Sel. Top. Quantum Electron. **6**(6), 1014–1021 (2000)
7. P.M. Nellen, A. Frank, R. Broennimann, U. Meier, U.J. Sennhauser, Fiber optical Bragg grating sensors embedded in CFRP wires, in *Smart Structures and Materials, Proceedings of SPIE*, vol. 3670 (1999)
8. M. Huang, Z. Zhou, Y. Huang, J. Ou, A distributed self-sensing FRP anchor rod with built-in optical fiber sensor. Measurement **46**(4), 1363–1370 (2013)
9. R.R.J. Maier, W.N. MacPherson, J.S. Barton, M. Carne, M. Swan, J.N. Sharma, Embedded fiber optic sensors within additive layer manufactured components. IEEE Sens. J. **13**(3), 969–979 (2013)
10. C.R. Farrar, K. Worden, An introduction to structural health monitoring. Phil. Trans. R. Soc. A. 365303–365315 (2006)
11. Y. Liu, S. Nayak, Structural health monitoring: state of the art and perspectives. JOM **64**, 789–792 (2012)
12. C. Kralovec, M. Schagerl, Review of structural health monitoring methods regarding a multi-sensor approach for damage assessment of metal and composite structures. Sensors **20**, 826 (2020)
13. O.V. Bashkov, R.V. Romashko, V.I. Zaikov, S.V. Panin, M.N. Bezruk, K. Khun, I.O. Bashkov, Detecting acoustic-emission signals with fiber-optic interference transducers. Russ. J. Nondestr. Test. **53**, 415–421 (2017)
14. J.O. Willberry, M. Papaelias, F.G. Franklyn, Structural health monitoring using fibre optic acoustic emission sensors. Sensors **20**(21), 6369 (2020)
15. F. Yu, Y. Okabe, Q. Wu, N. Shigeta, Fiber-optic sensor-based remote acoustic emission measurement of composites. Smart Mater. Struct. **25**(10), 105033 (2016)
16. O.V. Bashkov, A.A. Bryansky, M.N. Bezruk, T.A. Efimov, D.A. Bobruiko, E.A. Rassolov, R.V. Romashko, Registration of acoustic emission signals in polymer composite material during its destruction by means of embedded adaptive fiber-optic sensor. Bull. Russ. Acad. Sci. Phys. **86**(1), S13–S19 (2022)
17. O.V. Bashkov, R.V. Romashko, H. Khon, M.N. Bezruk, V.I. Zaikov, I.O. Bashkov, Registration of acoustic emission waves in anisotropic composite plates by fiber-optic sensors, in *Asia-Pacific Conference on Fundamental Problems of Opto- and Microelectronics*, vol. 11024 (SPIE, 2019), pp. 143–147

18. R.V. Romashko, T.A. Efimov, Y.N. Kulchin, Laser adaptive holographic system for microweighing of nanoobjects. Quantum Electron. **44**(3), 269 (2014)
19. J. Frejlich, A.A. Kamshilin, V.V. Kulikov, E.V. Mokrushina, Adaptive holographic interferometry using photorefractive crystals. Opt. Commun. **70**(2), 82–86 (1989)
20. R.V. Romashko, Y.N. Kulchin, M.N. Bezruk, S.A. Ermolaev, Laser adaptive holographic hydrophone. Quantum Electron. **46**(3), 277 (2016)
21. O.V. Bashkov, R.V. Romashko, I.O. Bashkov, V.I. Zaikov, H. Khon, D.B. Solovev, Sensitivity analysis of adaptive holographic fiber-optic acoustic emission sensors during the registration of acoustic waves in a plate. IOP Conf. Ser. Earth Environ. Sci. **459**(6), 062104 (2020)

Chapter 2
Physical Method for Determining the Parameters of a Hydraulically Satisfactory Reliable Cross-Section of Main Channels

Ahliman Verdiyev and Maya Abdullayeva

Abstract The article is devoted to the development of a physical method of simplified calculation of the live section of trapezoidal main channels of a satisfactory profile. The purpose of the study is to improve the method of determining the hydraulic parameters of the channel based on the hydraulically favorable profile of the section and the minimum limit value of the water filling depth. To do this, depending on the relative width from the bottom of the channel with different slope coefficient, the characteristics of its live cross-sectional area and average water flow were analyzed in comparison with the hydraulically most favorable cross-section profile. The change in the relative width, corresponding to the permissible decrease in the average flow rate and slope coefficient of the channel intervals, is analyzed. The hydraulic parameters of the semi-circular channel, accepted as a reference version of the comparison, were used to determine the level of suitability of the hydraulic parameters of the channel section for the given intervals. Since the semicircular channel is hydraulically more favorable than the trapezoidal channel under the same initial conditions, it has the smallest working section and the output parameters do not change. Also, the corresponding hydraulic parameters of the trapezoidal and semicircular channels calculated for the same flow rate were compared. From the point of view of construction technology and hydraulics, a simple calculation method for calculating hydraulic parameters has been developed for intervals corresponding to the relative width of the channel with a satisfactory cross-section profile from below. A method for calculating the parameters of a trapezoidal channel based on the hydraulic parameters of a semicircular channel is also proposed. On the basis of this method, knowing the hydraulic radius of the semicircular channel or the average speed of the water flow, using the corresponding ratio on the slope coefficient, it is

A. Verdiyev
"Water and Amelioration Scientific Research Institute", Public Legal Entity, Baku, Azerbaijan

M. Abdullayeva (✉)
Azerbaijan State Oil and Industry University, Baku, Azerbaijan
e-mail: mayaabdullayeva@hotmail.com

© The Author(s), under exclusive license to Springer Nature Switzerland AG 2025
A. Gibadullin et al. (eds.), *Proceedings of the 4th International Scientific and Practical Symposium on Materials Science and Physics*, Springer Proceedings in Physics 318,
https://doi.org/10.1007/978-3-031-94805-3_2

possible to easily calculate the satisfactory hydraulic parameters of the trapezoidal channel based on the proposed method.

2.1 Introduction

In practice, the geometric and hydraulic parameters (especially the relative bottom width, water depth corresponding to the maximum flow rate, velocity, hydraulic radius, etc.) of trapezoidal main earthen channel canals with the same water capacity and operated in different engineering and geological conditions differ sharply from each other. Although, on the one hand, this is due to engineering and geological conditions, on the other hand, it is tied to the methodology for determining the parameters of the live cross-section of the channel at the design stage, the need to take into account the level of accessibility of construction and other factors. There are many methods for designing open main earthen channel canals and reliable information about them is reflected in the relevant scientific works [1]. Based on these methods, by performing many painstaking calculations with successive approximations, the required parameters of the live cross-section of the channel are determined [2]. Therefore, the solution to the problem of easy determination of optimal and hydraulically satisfactory parameters of the channel cross-section, in accordance with specific geological conditions, construction capabilities and efficient land use has not yet been sufficiently studied and constantly attracts the attention of researchers. During the design period, the parameters of the cross-section profile of the main earthen channel channels are determined taking into account the stability of the slopes, non-eroding and non-silting flow rates, as well as the level of accessibility of construction. However, during long-term operation, the cross-section parameters of earthen channels are subject to fundamental changes. In this regard, in order to increase reliability, it is necessary to predict the functional parameters of the channels during the design period, using various forecasting methods, including the Monte Carlo method [3]. Since the stability of the slopes of earthen channels depends on soil conditions, the issues of hybrid optimization for calculating a stable and optimal channel cross-section, taking into account their stability by the slope coefficient [4], taking into account channel deformations for a stable channel when restoring capacity associated with flushing during operation [5] attract the attention of researchers. In practice, when designing open channels, the identification of the optimal hydraulic cross-section for various cross-sectional shapes [6–8] as well as the stable and optimal hydraulic cross-section of a trapezoidal channel under certain limiting conditions [9, 10], the optimal design of channels with a parabolic cross-section with a horizontal bottom [11] and with a curved bottom [12] with maximum water throughput, based on a mathematical model, the possibility of calculating channels with different cross-sectional shapes with a steady flow in open channels [13] have been studied. The parameters of the hydraulically most advantageous channel section were identified by studies conducted by Uginchus [14] and Latyshenkov [15], and in order to ensure the availability of construction conditions, it was proposed to replace the design hydraulically most advantageous cross-section, with the corresponding average velocity, with a live section profile differing by 2–3% from the average velocity. Kosichenko [16, 17] investigated the hydraulic parameters of polygonal and trapezoidal channels, and Mikhnevich [18] studied the hydraulic parameters of trapezoidal channels; accordingly, methods were developed for determining the most advantageous cross-section of a polygonal [19] and trapezoidal profile [20], and the proposal of Latyshenko [15] was taken as a basis for the transition to a cross-section close to the hydraulically most advantageous profile. Based on the proposed methods for determining the parameters of the live section of a trapezoidal channel taking into account the slope coefficient, various values of the geometric and hydraulic parameters of the channel were obtained depending on the choice of a suitable value of the relative width of its bottom. However, it remains unclear which method can be used to calculate the hydraulically more efficient parameters of the cross section. In this regard, when designing channels, along with the same transport capacity, the hydraulically most advantageous cross-sectional profile with the smallest cross-sectional area is selected as a comparison option, and the idea of conducting an appropriate assessment based on a comparison with this profile is put forward. Since the hydraulically most advantageous profile with the smallest cross-sectional area is a semicircular section, an assessment of the level of profitability of the design option of a trapezoidal channel can be made.

2.2 Purpose and Methodology

It consists in the improvement of the method of determining the hydraulic parameters of the channel according to the hydraulically favorable cross-sectional profile area and the minimum limit value of the water filling depth.

The consumption of a trapezoidal main channel is calculated using the following appropriate Shezy formula for the known cross-sectional area and the flow rate during uniform fluid movement [21]:

$$\begin{cases} Q = \omega \cdot \vartheta; \\ \vartheta = C \cdot \sqrt{R \cdot J} = K \cdot \sqrt{J}; \\ C = \frac{1}{n} \cdot R^{y}; \\ y = 2.5 \cdot \sqrt{n} - 0.13 - 0.75 \cdot \sqrt{R} \cdot \left(\sqrt{n} - 0.10\right); \\ K = \frac{Q}{\sqrt{J}} = \omega \cdot C \cdot \sqrt{R}. \end{cases} \tag{2.1}$$

Here:

Q channel consumption i, m^3/s;
ω live cross-sectional area of the flow, m^2;
ϑ average flow rate calculated by Shezi's formula, m/s;
C Shezi coefficient, $\frac{m^{0.5}}{san}$;

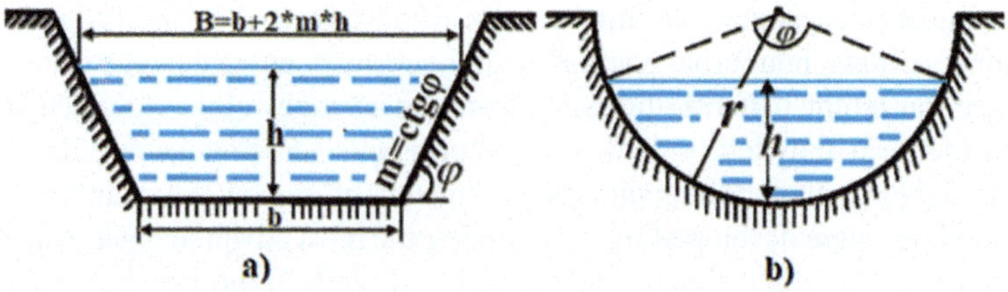

Fig. 2.1 Cross-sectional profiles of the channel: **a** trapezoidal; **b** semicircular

R hydraulic radius, m;
J the slope of the water surface of the stream;
K consumption module, m^3/s;
y Shezi coefficient, the superpower index determined by the formula of Pavlovsky [21];
n roughness coefficient of the channel.

In practice, when designing trapezoidal trunk channels (Fig. 2.1a), their live section parameters are calculated by the following formulas.

$$\begin{cases} \omega = b \cdot h + m \cdot h^2 = (\beta + m) \cdot h^2; \\ \chi = b + 2h \cdot \sqrt{1+m^2} = \left(\beta + 2 \cdot \sqrt{1+m^2}\right) \cdot h; \\ R = \frac{\omega}{\chi} = \frac{(\beta+m)\cdot h}{\left(\beta+2\cdot\sqrt{1+m^2}\right)}. \end{cases} \tag{2.2}$$

Here:

ω live cross-sectional area of the channel, m^2;
b channel bottom width, m;
h water depth in a trapezoidal channel, m;
m slope coefficient of the channel ($m = ctg\varphi$);
β the relative width of the channel from the bottom ($\beta = \frac{b}{h}$);
χ channel wetting perimeter, m;
R hydraulic radius of the channel, m.

Relative width from the bottom for the hydraulically most favorable cross-sectional profile of the channel [22], is calculated by the following formula.

$$\beta_{h.a.} = 2 \cdot \left(\sqrt{1+m^2} - m\right). \tag{2.3}$$

Here:

$\beta_{h.a.}$ relative width from the bottom on the hydraulically most favorable cross-sectional profile of the channel;
m channel slope coefficient, m = ctgφ.

If the subflow capacity (consumption) of the channel, the roughness coefficient of the channel, the hydraulic gradient are known in advance, then the hydraulic parameters can be calculated with some simplification using the consumption formula (2.1) for uniform movement of water in the channel. In this case, the radius of the channel with a semicircular cross-section profile (Fig. 2.1b) (for the case when $\varphi = \pi$, $\omega = \frac{\pi \cdot r^2}{2}$, $\chi = \pi \cdot r$, $R = \frac{r}{2}$) and trapezoidal channel. formula of water depth, corresponding to permeability, are expressed as follows:

$$\begin{cases} r = \left(\frac{2^{1.5+y} \cdot K \cdot n}{\pi}\right)^{\frac{1}{2.5+y}}; \\ h = \left[\frac{\left(K \cdot n \cdot \left(\beta + 2 \cdot \sqrt{1+m^2}\right)^{0.5+y}\right)}{(\beta+m)^{1.5+y}}\right]^{\frac{1}{2.5+y}}. \end{cases} \quad (2.4)$$

Here:

- r radius semicircular section, m;
- K consumption module, m^3/s;
- n roughness coefficient of the channel;
- y Shezi coefficient, the superpower index determined by the formula of N. N. Pavlovsky;
- π 3.14;
- h depth of water corresponding to the subflowing capacity of the channel, m;
- β the relative width of the channel from the bottom, i.e. $\beta = \frac{b}{h}$;
- b channel bottom width, m.

At the hydraulic radius R = 0.5–3.0, researchers [23], proposed to accept approximately y = 0.2. According to the conducted analyses, the value of y decreases as the value of the roughness coefficient decreases at fixed values of the hydraulic radius, and the value y increases as the value of the roughness coefficient increases. When designing trapezoidal open channels, the roughness coefficient in most cases is accepted within n = 0.014–0.022, and at this time it varies within y = 0.15–0.20. Therefore [24], y ≈ 0.17 can be taken according to Manning's formula R (1891) for trunk channels in the first approximation, then:

$$\begin{cases} r = \left(\frac{2^{1.67} \cdot K \cdot n}{\pi}\right)^{\frac{1}{2.67}}; \\ h = \left[\frac{\left(K \cdot n \cdot \left(\beta + 2 \cdot \sqrt{1+m^2}\right)^{0.67}\right)}{(\beta+m)^{1.67}}\right]^{\frac{1}{2.67}}. \end{cases} \quad (2.5)$$

maybe. In this case, he calculated according to the formula N. N. Pavlovsky, according to the given initial parameters, it is possible to set the values of the Shezi coefficient, speed and flow rate.

At the hydraulically most favorable cross-sectional profile of a trapezoidal channel, the cross-sectional area is small, but the water depth is large. This creates a corresponding difficulty for the construction and operation period [22], Therefore, the cross-sectional area of the channel and the water depth should be as small as possible. The reduction of the cross-sectional area of the channel and the water depth can be carried out in such a way that the ratio of the water velocity in the channel to the water velocity on the hydraulically most favorable cross-section profile is equal to 0.97. By maintaining this ratio, a relationship can be established between the parameters of the trapezoidal channel and the parameters of the semicircular channel. When such a relationship is established analytically, the parameters of the required cross-sectional profile of the trapezoidal channel can be calculated based on the parameters of the semicircular channel. Therefore, at fixed values of the subflow capacity of the trapezoidal channel, the change of the flow rate, the cross-sectional area of the channel, and the water depth should be analyzed depending on the change of its relative width from the bottom. In this case, the water flow rate should be taken as the main parameter and compared with the flow rate corresponding to the hydraulically most favorable cross-sectional profile (in trapezoidal and semicircular sections). If the central angle $\varphi = \pi$, the live cross-sectional area $\omega = \frac{\pi \cdot r^2}{2}$, the wetting perimeter $\chi = \pi \cdot r$ and the hydraulic radius $R = r/2$, the parameters of the channel with the semicircular cross-sectional profile calculated by the formulas are the minimum cross-section It is known that it is suitable for area and hydraulics [25, 26]. Therefore, the parameters determined by the proposed method can be compared based on the hydraulic parameters of this profile.

2.3 Discussion and Results

Using formula (2.5) based on any accepted values of the relative width of the channel from the bottom, it is possible to calculate the water depth corresponding to its flow rate. Let us assume that the flow rate of the designed trapezoidal open main channel is $Q = 10\ m^3/s$, the slope coefficient $m = 1.5$, the roughness coefficient $n = 0.020$, the hydraulic slope $J = 0.0001$. It is necessary to determine the hydraulic parameters for the case when the relative width of the channel from below is equal to $\beta = 0.99$.

If these ratios are calculated comparatively for the live cross-sectional area and mean flow velocity of the most favorable cross-sectional profile of a trapezoidal channel instead of a semicircular profile, then the ratios 16.13/16.10 = 1.002 and 0.624/0.625 = 0.998 are close to one.

On the example of a trapezoidal channel with a discharge capacity of $Q = 10\ m^3/s$, the change of the live cross-sectional area and the average flow velocity (in different slope coefficients) depending on its relative width from the bottom was analyzed, and the results of those analyzes were graphically depicted (Fig. 2.2). Based on the graphical representation given in Fig. 2.2, it can be said that as the relative width of the channel from the bottom increases from zero, its live cross-sectional area first decreases and takes the smallest value at the relative width from the bottom corresponding to the most favorable cross-sectional profile ($\beta_{h.a.}$), then, as β increases, the live cross-sectional area also increases. As β increases, the average water flow velocity first increases and then decreases. The average flow velocity takes a maximum value when the relative width of the channel from the bottom is equal to the relative width from the bottom for the most favorable cross-sectional profile. At the same time, as the slope coefficient of the channel increases, its cross-sectional area increases, and the average flow velocity decreases. The smallest live cross-sectional area and the largest average flow velocity are obtained for the semi-circular profile, while the matching conditions remain the same. The hydraulic parameters of the trapezoidal channel with the same flow capacity on the most favorable cross-sectional profile were compared with the parameters of the semicircular channel, Table 2.1 was compiled based on the obtained results.

Knowing the hydraulic radius of the semicircular channel, the ratio given in column 5 of Table 2.1 can be used to calculate the hydraulic radius of the trapezoidal channel with the most hydraulically favorable profile and its other parameters.

According to the Shezi formula, the average flow rate of water on cross-sectional profiles (semicircular and trapezoidal) with the same subflow capacity of the channel:

$$\vartheta_1 = \frac{1}{n} \cdot R_1^{y+0.5} \cdot \sqrt{J}; \quad \vartheta_2 = \frac{1}{n} \cdot R_2^{y+0.5} \cdot \sqrt{J} \tag{2.6}$$

since the ratio of average flow velocities is as follows:

$$\frac{\vartheta_1}{\vartheta_2} = \left(\frac{R_1}{R_2}\right)^{y+0.5}. \tag{2.7}$$

Here:

- ϑ_1 and ϑ_2 respectively, average flow velocities of water on the trapezoidal and semicircular cross-section profile, m/s;
- R_1 and R_2 hydraulic radii on trapezoidal and semicircular cross-sectional profiles, respectively, m;
- y power index determined by N. N. Pavlovsky's formula for calculating the Shezi coefficient.

According to the formulas (2.1) and (2.6), it can be said that the hydraulic radius and the average flow velocity on the semicircular profile of the channel are unchanged for constant values of the roughness coefficient, hydraulic slope and flow. The hydraulic radius and average flow velocity of a trapezoidal channel with a suitable slope coefficient will also vary depending on the relative bottom width (bottom width and water depth). Therefore, the ratio (2.7) will be constant for each value of the relative width from the bottom of the trapezoidal channel with the corresponding slope coefficient, if the roughness coefficient, hydraulic slope and flow do not change. In this regard, the parameters calculated for the semicircular profile can be used to easily calculate the parameters of the trapezoidal cross-section capable of releasing the required flow of the channel.

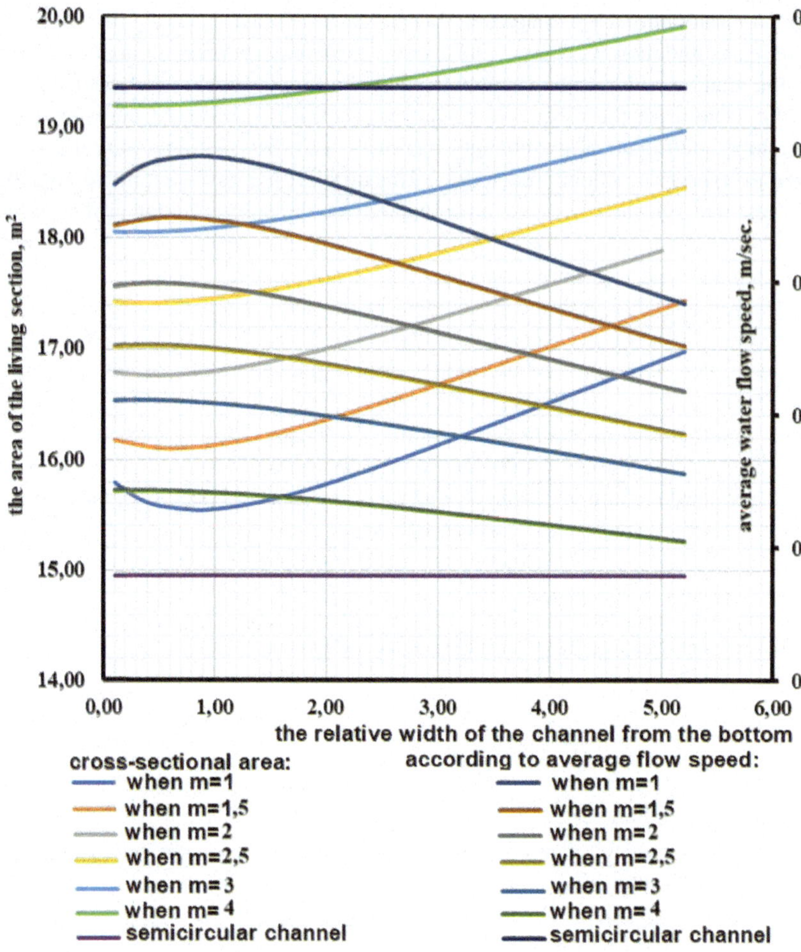

Fig. 2.2 Variation of the live cross-sectional area and average flow velocity depending on the relative width of the trapezoidal channel from the bottom at different slope ratios (when Q = 10 m^3/s)

The decrease of the average flow rate of water occurs gradually according to the change of the relative width of the channel from the bottom. According to its slope coefficients, the interval of change of the relative width from the bottom according to the value $A_\vartheta = 0.97$ was determined. Depending on the slope coefficient corresponding to those intervals, the ratios of the parameters of the trapezoidal channel (average flow velocity and hydraulic radius) to the corresponding parameters of the semicircular channel were also calculated and Table 2.2 was compiled based on the obtained results.

During the design of underground channels, the following condition required for the average flow velocity must be satisfied:

Table 2.1 Comporison of hydralic radis for various cross-sectional profiles of channels

Slope coefficient	Relative botton width of a trapeoidal channel according to the most avantaeos hydalic cross-section profile, β_{rb}	Hydraulic radis		Ratio, $\frac{R_{h.r.}}{R_r}$
		A trapezoidal channel according to the most advantageous hyraulic cross-section profile	A semi-circular channel according to the cross-section profile	
1.00	0.828	1.458	1.543	0.945
1.50	0.606	1.383	1.543	0.896
2.00	0.472	1.302	1.543	0.844
2.50	0.385	1.229	1.543	0.796
3.00	0.325	1.165	1.543	0.755
4.00	0.246	1.063	1.543	0.689

Table 2.2 Comporison of parameters for the channels of a semicircular transverse profile and trapeoidal with a transverse profile close to the hydraulically most advatageous

Slope coefficient, m	Elative botton width of the channel, β	$A_\vartheta = \frac{\vartheta}{\vartheta_{h.\partial.}}$	The ratio of the parameters of the trapezoidal channel to the corresponding parameters of the semi-circlar channel	
			Based on the average flow speed, $\frac{\vartheta}{\vartheta_r}$	Based on the hydraulic radius, $\frac{R}{R_r}$
1.00	2.66–2.69	0.970	0.934	0.904
1.50	2.71–2.75	0.970	0.900	0.857
2.00	2.94–2.99	0.970	0.864	0.807
2.50	3.29–3.29	0.970	0.830	0.762
3.00	3.59–3.66	0.970	0.800	0.722
4.00	4.42–4.49	0.970	0.752	0.660

$$\vartheta_l < \vartheta < \vartheta_y. \tag{2.8}$$

Thus, if the speed ϑ according to condition (2.8) is known, then the parameters of the trapezoidal channel can be determined in the following order:

Option I. The flow rate of the channel is Q = 10.00 m^3/s, slope coefficient m = 1.5, roughness coefficient $n = 0.020$ hydraulic slope J = 0.0001, average flow velocity $\vartheta = 0.610$ m/s.

Option II. Calculation of trapezoidal channel parameters using semicircular channel parameters. Assume that the flow of the channel is $Q = 100.00$ m^3/s, the slope coefficient $m = 1.5$, the roughness coefficient n = 0.020, the hydraulic gradient $J = 0.0001$, $K = 10{,}000.00$ m^3/s given.

Option III. Calculation of hydraulic parameters based on the relative width of the trapezoidal channel from the bottom. Assume that the flow of the channel is $Q = 100.00\ m^3/s$, the slope coefficient $m = 1.5$, the roughness coefficient $n = 0.020$, the hydraulic gradient $J = 0.0001$, $K = 10{,}000.00\ m^3/s$ given.

Thus, it is possible to calculate the hydraulic parameters of the underground channel in the appropriate sequence based on the required initial data.

2.4 Conclusion

Based on the results of the study, the following conclusions can be drawn:

- Parameters corresponding to the cross-sectional profile of the semi-circular channel, which is the most hydraulically effective, can be adopted as a reference option when determining the hydraulic efficiency of other profiles of the channel;
- The hydraulic parameters for the cross-sectional profile of the trapezoidal channel can be calculated based on the hydraulic parameters (hydraulic radius and average flow velocity) of the cross-sectional profile of the semicircular channel corresponding to the same initial condition;
- On the basis of the proposed physical methodology, the hydraulic report of the trapezoidal canals, which are convenient from the point of view of hydraulics and construction, can be quickly carried out.

References

1. V.T. Chow, *Open-Channel Hydraulics* (McGraw-Hill Book Company, New York, 1959), 692 pp.
2. E.I. Mikhnevich, Calculation of the capacity and stability of channels. Ecol. Constr. **1**, 23–31 (2020). https://doi.org/10.35688/2413-8452-2020-01-003
3. A.A. Verdiev, Methodology for predicting the confidence interval for the water capacity of channels at the design stage. Bull. Sci. Pract. **8**(12), 372–381 (2024). https://doi.org/10.33619/2414-2948/85/44
4. R.K. Bhattacharjya, M.G. Satish, Optimal design of a stable trapezoidal channel section using hybrid optimization technique. J. Irrig. Drain. Eng. **133**, 323–329 (2007)
5. A.B. Saparov, S.K. Khidirov, A.B. Khalimbetov, A.T. Djumanazarova, Sh.B. Tolepova, A.R. Seitmuratov, Methods of channel calculations to improve the operational mode of main canals. E3S Web Conf. **410**, 05035 (2023). https://doi.org/10.1051/e3sconf/202341005035
6. S.O. Kurbanov, N.V. Khanov, Fundamentals of optimization of polygonal sections of hydraulic channels. Hydrotech. Constr. **12**, 27–31 (2008)
7. G. Sun, W. Wei, P. Zhang, G. Jang, W. Hu, Calculation of optimal hydraulic cross section of a cubic parabola-shape open channel, in *Asia-Pacific Power and Energy Engineering Conference*, 5748593 (2011)
8. S.O. Kurbanov, N.V. Khanov, On the hydraulic calculation of the most advantageous sections of energy channels of polygonal profile. Hydrotech. Constr. **7**, 40–43 (2003)

9. M.A. Mikhalev, Calculation of main channels. St. Petersb. Eng. Constr. J. **4**, 83–93 (2013). https://doi.org/10.5862/MCE.39.9
10. A. Das, Chance constrained optimal design of trapezoidal channels. J. Water Resour. Plan. Manag. **134**, 310–313 (2008)
11. Y. Han, X. Gao, Z. Xu, The best hydraulic section of horizontal-bottomed parabolic channel section. J. Hydrodyn. **29**(2), 305–313 (2017). https://doi.org/10.1016/S1001-6058(16)60740-0
12. I. Sarbu, A. Iosif, A general analytical model for hydraulic computation of open channels with steady state uniform flow. Int. J. Adv. Appl. Sci. **5**(1), 1–7 (2018)
13. B.R. Chahar, Optimal design of a special class of curvilinear bottomed channel section. J. Hydraul. Eng. **133**(5), 571–576 (2007)
14. A.A. Uginchus, *Hydraulic and Technical-Economic Calculations of Channels* (Stroyizdat, Moscow, 1965), 274 pp.
15. A.M. Latyshenkov, *Canals of Water Supply and Irrigation Systems* (Stroyizdat, Moscow, 1972), 153 pp.
16. Yu.M. Kosichenko, O.A. Baev, Hydraulic efficiency of irrigation canals during operation. Bull. MGSU **15**(8), 1147–1162 (2020). https://doi.org/10.22227/1997-0935.2020.8.1147-1162
17. Yu.M. Kosichenko, E.G. Ugrovatova, Improving the efficiency of operation of large canals and substantiation of the shape and hydraulic resistance of polygonal cross-section channels. News of higher educational institutions. North Caucasus region. Techn. Sci. **2**(198), 96–103 (2018). https://doi.org/10.17213/0321-2653-2018-2-96-103
18. E.I. Mikhnevich, *Open Watercourses: Capacity and Stability* (BNTU, Minsk, 2021), 311 pp.
19. Yu.M. Kosichenko, Study of the hydraulically advantageous profile of the polygonal section of large channels and their hydraulic resistances. Nat. Manage. **2**, 85–89 (2014)
20. E.I. Mikhnevich, Optimization of the parameters of main channels in the area of the most hydraulically advantageous cross-sections. Sci. Technol. **22**(3), 208–215 (2023). https://doi.org/10.21122/2227-1031-2023-22-3-208-215
21. P.G. Kiselev, *Handbook of Hydraulic Calculations* (Energiya, Moscow, 1972), 312 pp.
22. N.N. Pavlovsky, *Hydraulic Handbook* (Main Editorial Office of Energy Literature, Moscow, 1937), 890 pp.
23. I.I. Agroskin, *Hydraulics* (Energy, Moscow, 1964), 352 pp.
24. R. Manning, On the flow of water in open channels and pipes. Proc. Inst. Civ. Eng. Ireland **20**, 161–206 (1890)
25. W.H. Graf, *Fluvial Hydraulics* (Wiley, England, 1998), 681 pp.
26. A.R. Vatankhah, Semi-regular polygon as the best hydraulic section in practice (generalized solutions). Flow Meas. Instrum. **38**, 67–71 (2014). https://doi.org/10.1016/j.flowmeasinst.2014.05.016

Chapter 3
Physical Principles of Using New Promising Materials and Technologies for Environmentally Friendly and Safe Nuclear Power Reactors

V. S. Okunev

Abstract The article discusses new materials for high-power fast nuclear power reactors. Among promising materials, nanomaterials, composite materials, and powder materials are considered. The physical principles of using new materials are analyzed. Nanomaterials and nanopowders can be used in the production of nuclear fuel and as additives to structural materials of fuel pin cladding. The article discusses reactors with liquid lead coolant. These reactors will form the basis of large-scale nuclear power in Russia, the transition to which is predicted to be by the middle of the twenty-first century. To improve the safety of reactors, it is proposed to use innovative cermets fuel with additives of enriched or depleted uranium nanopowder. It is proposed to use lead extracted from thorium ores as a coolant. A reactor with an electrical capacity of up to 3 GW is characterized by negative void and density reactivity effects under the most dangerous scenario of their implementation. It is resistant to any emergency situations, including anticipated transient without scram. The proposed innovations (new materials of the core) will allow creating a high-temperature reactor, including an abnormally high power. The article examines chemical reactions of free hydrogen formation in heat exchangers and the core, which can be realized in an accident with contact of lead and steel components with superheated water vapor of the second circuit. A hydrogen explosion is possible only upon contact with atmospheric air during a hypothetical destruction of the reactor or heat exchanger. This can lead to aggravation of the accident.

V. S. Okunev (✉)
Bauman Moscow State Technical University, Moscow, Russia
e-mail: okunevvs@bmstu.ru

© The Author(s), under exclusive license to Springer Nature Switzerland AG 2025

A. Gibadullin et al. (eds.), *Proceedings of the 4th International Scientific and Practical Symposium on Materials Science and Physics*, Springer Proceedings in Physics 318,
https://doi.org/10.1007/978-3-031-94805-3_3

3.1 Introduction

The physical and chemical laws on which the technologies of the near future can be based are well known. The main problem lies directly in the technologies and their development. The development of new technologies of the future is based on the development and use of innovative materials with specified properties. Scientists are increasingly introducing nanomaterials and nanotechnologies, powder materials, composites and composite technologies (including nanopowder, micro powder and other technologies) into production, combining dissimilar materials into a common structure, the properties of which differ greatly from the properties of individual components.

The predicted growth of energy consumption requires the search for new energy sources that are safe, cost-effective, and have a long-term supply of fuel. Such energy sources include systems based on the transmutation of atomic nuclei. These are reactors for the fission of heavy nuclei and reactors for the fusion of light nuclei.

The processes of synthesis of light nuclei into heavier ones and decay of heavy nuclei into lighter ones are energetically favorable. The directions of spontaneous evolution in nature are shown by arrows in Fig. 3.1.

The maximum in Fig. 3.1 corresponds to mass numbers A $\approx$ 50–60. The highest average specific binding energy is in the nuclei of ^{56}Fe and ^{62}Ni. These are the so-called average nuclei. Spontaneous evolution in nature is moving towards the "average": the future belongs to the golden mean, or "yellow", the middle color of the five-color rainbow in Chinese philosophy. Light nuclei tend to move into the category of heavier ones, heavy nuclei tend to move into the category of lighter (average). For this reason, it became possible to obtain energy in fission reactors of heavy nuclei, in radioactive decays of heavy nuclei, and in installations for the synthesis of light nuclei. The use of such installations, i.e. obtaining energy as a result of the transmutation of atomic nuclei, significantly accelerates evolutionary processes in nature and is a practically irreversible process.

Heavy nuclear fission reactors are much more technologically advanced than fusion reactors. As of November 2, 2024, there were 415 nuclear power units in operation worldwide with a total installed electric capacity (net) of about 373,735 MW, and 63 nuclear power units with an electric capacity of 66,100 MW were under

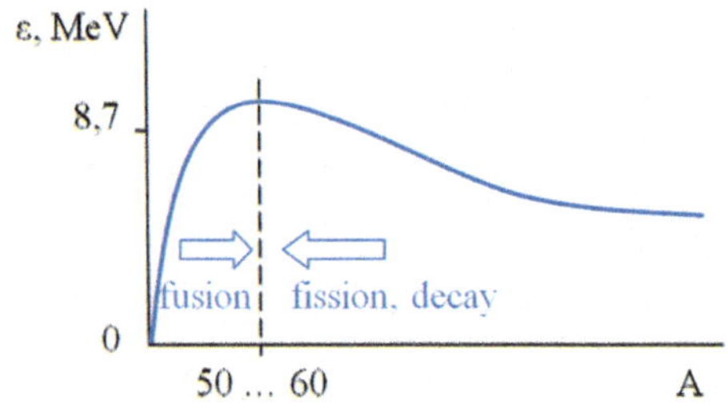

Fig. 3.1 Dependence of the average specific energy ε of the nuclear bond on the mass number A and the direction of spontaneous evolution in nature (indicated by arrows)

construction [1]. The world nuclear power industry has generated 20,053 reactor-years [1, 2]. The fuel base of nuclear fission reactors is sufficiently broad to use them as the main energy source in the long term. In addition to traditional uranium fuel (the possibility of expanded reproduction of fissile ^{239}Pu, disposal of weapons-grade plutonium), huge reserves of thorium (raw material for the production of fissile ^{233}U) can be considered as a solution to the fuel supply problem.

There are about 90 fusion reactors operating worldwide. According to a December 2022 report by the International Atomic Energy Agency (IAEA), more than 130 experimental fusion reactors, both public and private, are operating, under construction, or planned in 50 countries [3].

Heavy nuclear fission reactors are potentially dangerous. During the operation of the reactors, there have been four serious accidents: at the Windscale reactor (United Kingdom, October 10, 1957); at the Three Mile Island nuclear power plant (USA, 1979); at the Chernobyl nuclear power plant (USSR, April 26, 1986); and at the Fukushima I nuclear power plant (Japan, March 11, 2011).

The specificity of nuclear energy is that almost all of its waste is radioactive. (To a certain extent, this is compensated by the small volume of waste: no more than 5–7% of the fuel volume.)

The safety of thermonuclear reactors is mainly related to instability, disruption or termination of plasma combustion in any emergency situations. In this case, the reactor stops working. Thermonuclear reactor waste, unlike nuclear power plant waste, has a relatively low specific radioactivity and does not contain transuranic elements, fission products or radioactive gases (except tritium). Thermonuclear installations do not contain materials that require cooling for decades.

It should be noted that modern nuclear power plants with fission reactors can only partially meet the world's energy needs. The fuel base of nuclear power with such reactors is broad, but still limited (even with the expanded reproduction of fissile materials), since natural radioactive elements (uranium and thorium) are not very common, their reserves in nature are limited, and deposits are unevenly distributed. As a result, many countries face the problem of importing nuclear fuel.

With regard to the inexhaustibility of the fuel base, it should be noted that in the foreseeable future, thermonuclear reactors are focused on the reaction ^{2}H + ^{3}H. The limiting factor in the fuel base is lithium, the raw material for the production of tritium. Lithium is an element scattered in the earth's crust, the content of which to a depth of 16 km averages 0.64×10^{-4} g ^{6}Li per 1 g of rock (for comparison: the content of uranium is 0.36×10^{-5} g per 1 g of rock). Each gram of the earth's crust potentially contains 1.2×10^6 J/g of fusion energy and 3×10^5 J/g of fission energy of uranium, or 106 J/g of fission energy of uranium and thorium. (We are talking about reserves, most of which are inaccessible.) This means that the potential reserves of fusion and fission energy on Earth are of the same order of magnitude.

In the development of human society and technology, certain milestones are identified. In relation to the development of nuclear energy in Russia, the year 2050 was identified as a certain milestone at the beginning of the twenty-first century. It is defined as the end of the transition to large-scale nuclear energy in Russia [4].

Recently, this milestone in the development of world energy, dated 2050, has been considered by many domestic and foreign analysts [5].

Several trends in the development of global energy can be identified. One of the main trends is the reduction of the carbon footprint of the global fuel and energy complex by abandoning the priority in the use of organic fuel ("decarbonisation" of energy, achieving carbon neutrality of the global fuel and energy complex by 2050). Another trend is associated with the uncertainty of the dynamics of energy markets. The range of expert assessments presented in various forecasts is extremely wide. According to forecasts, natural gas consumption in 2050 will amount to from 500 billion m^3 to almost 8 trillion m^3 [4]. The share of renewable energy sources, including solar, wind, hydroelectric power plants and biomass installations, varies from 13 to 84% [4]. These trends formed the basis of the "World Energy Development Scenarios to 2050". The scenarios were prepared by a working group consisting of employees of the Russian Energy Agency of the Ministry of Energy of Russia and external experts [4].

Achieving carbon neutrality of the global fuel and energy complex by 2050 is an ideal scenario that is unlikely to be achieved. It is too expensive (capital-intensive) a way to transform the structure of the existing global energy sector. In the absence of financial assistance from economically developed countries, it is unacceptable for countries with poorly developed economies. "Decarbonization" of energy may be achieved to the detriment of other important socio-economic objectives of global development, regulated by the UN. In particular, one of the 17 goals of sustainable development—"affordable and clean energy" implies ensuring universal access to reliable, sustainable and modern energy sources [6].

Achieving carbon neutrality can be facilitated, for example, by a significant increase in the scale of nuclear energy development. There is no other alternative in sight yet. Forecasts for electricity generation at nuclear power plants speak in favor of the transition to large-scale nuclear energy not only in Russia, but also in the world. If in Russia the share of electricity generation at nuclear power plants in the period from 2000 to 2050 will increase according to forecasts by 10 times, then in the world by two to three times [4, 5]. The maximum coal consumption in the global electric power industry was passed in 2020 [4]. The maximum consumption of natural gas falls on 2030 [4, 5]. In subsequent years, a decline in the consumption of organic fuel is predicted.

There are ten operating nuclear power plants in Russia, with 33 power units [1, 2]. According to Rosatom State Corporation, Russian nuclear power plants generate 25,242 MW of electricity. The leaders in energy production at nuclear power plants are the United States (102,709 MW), France (65,880 MW) and Japan (46,292 MW). Russia is in 8th place. The share of electricity generated by nuclear power plants in Russia is about 20% of all electricity produced. At the same time, in the European part of the country, the share of nuclear energy reaches 30%, and in the North-West—37%. Energy saving policy allows saving about 20% of electricity. Thus, at present, energy saving will allow abandoning nuclear power plants in Russia.

The problem of creating environmentally friendly and safe nuclear power reactors of the new generation is relevant. New materials are needed for such reactors.

One of the most promising reactors of the new generation can be considered BREST-OD-300. The experimental demonstration power unit with the BREST-OD-300 reactor is being built at JSC Siberian Chemical Plant (Seversk, Tomsk region, Russia) [7]. The physical start-up of the reactor is scheduled for 2026. Severe accidents are excluded in the BREST-OD-300 reactor. ATWS (anticipated transient without scram) emergency situations do not lead to an accident. However, increasing the reactor power will require the use of new materials for the core to ensure safety.

3.2 Materials

3.2.1 Nanotechnology and Nanomaterials. Possibility of Improving the Properties of Materials

A qualitative leap in improving the physical, chemical and, ultimately, operational properties of materials (including those used in nuclear technology) is possible with the transition from micro- to nanotechnology of their manufacture.

Qualitatively new physical and chemical properties of nanomaterials are explained by two factors associated with the reduction in the size of the object (nanoobject):

- The ratio of the number of atoms on the surface to the number of atoms in the volume of the object increases. The influence of surface interaction forces and interfaces on the properties of the object grows. The result is high chemical activity of nanomaterials.
- The quantum properties of the object are manifested to a greater extent. The uniqueness of nanomaterials is associated with the internal energy density of the material, determined by the crystalline structure [8]. The higher the energy density, the larger the permissible particle sizes. With a decrease in the size of nanostructures, the time of various processes decreases, and the role of the tunnel effect increases [9]. The upper limit of sizes separating nanoobjects from microobjects is conditional and is determined by the dominance of the laws of quantum physics in nanoobjects.

There are two main directions of nanotechnology development.

At the first stage of nanotechnology development, materials are ground to nanosized powder and used as one of the components of a composite material. For example, in our country, nanopowders have been used in nuclear technology since the 1960s under the name of fine powders. In general, fine powders are powders with a characteristic particle size of 100 nm–10 μm. Nanoparticles are particles with characteristic sizes from 1 to 100 nm.

Another, more fundamental direction is connected with the development of methods of production and application of products with a given atomic structure by means of controlled manipulation of individual atoms and molecules. The nearest prospects of development of nuclear power engineering are connected only with the first direction.

The term "nanotechnology" was first used by Japanese physicist Norio Taniguchi in 1974. He used this term to describe the processes of creating semiconductor structures with nanometer-level accuracy. "Nanotechnology primarily consists of processes of separating, combining, and deforming materials atom by atom or molecule by molecule" [10]. Later, this term was used by engineer Eric K. Drexler [11, 12].

In ISO/TC 229, nanotechnology means knowledge and control of processes, typically on a scale of 1 nm, but not excluding a scale of less than 100 nm, in one or more dimensions, where the exploitation of size effects (phenomena) leads to the possibility of new applications; the use of properties of objects and materials at the nanometer scale that differ from the properties of free atoms or molecules, as well as from the bulk properties of matter consisting of these atoms or molecules, to create more advanced materials, devices, systems that implement these properties [13].

Nanoscience studies the fundamental properties of nanometer-sized objects (nanoobjects) and related phenomena. Physics and technology of nanostructures, atomic and molecular physics is an area of fundamental and applied science and technology that includes studies of the physical properties of objects with characteristic dimensions of ~ 100 nm and less, at least in one direction, as well as systems of such objects for which key characteristics are determined by the size of the system.

Modern development of nanotechnology began with the discovery of three carbon nanostructures: fullerenes (from C60 to C100)—spherical molecules, nanotubes—monatomic layers of graphite rolled into a cylinder with an external diameter of 1–5 nm, graphene—a monatomic layer of graphite. Fullerene C60 was discovered in 1985, carbon nanotubes—in 1991, graphene was obtained in 2004. The first publication on the observation of fibers with a diameter of about 100 nm was published in 1952 [14]. Nanotubes were observed in 1953 [15].

Research by P. Paufler (Dresden University of Technology, Germany) has shown that the surface of a sixteenth-century Damascus blade contains carbon nanotubes (with a layer spacing of 0.34 nm) and cementite Fe_3C nanofibers [16]. The patterns on crucible damask steel blades (wootz steel) are the result of the formation of large Fe_3C matrices on the steel surface with a high carbon content (up to 1.5–2%) and slow cooling. Indian crucible steel was first mentioned around 300 BC. Aristotle mentioned crucible smelting of metals in the fourth century BC [17].

Main areas of nanotechnology application. Nanotechnology is used to obtain structural materials and coatings with specified characteristics. Optimization of material properties is usually achieved by alloying and deformation. In the field of nanotechnology, optimization of material properties is achieved by adding nanomaterials (ultra dispersed elements) as alloying additives and/or creating ultra dispersed systems of non-metallic inclusions in steels and alloys, carried out through thermoplastic, thermal or plastic deformation. Deformation produces materials whose strength and heat resistance are 4 times higher than those used in mechanical engineering, and nanostructured steels are tens of times stronger than conventional steels [8]. Active minerals with shape and action memory are used to create coatings [8].

In mechanical engineering and power engineering, nanotechnology is used to optimize the properties of fuel based on micro- and nanoadditives of concentrators around which the combustion process occurs. The use of fullerenes is hampered by their high cost (hundreds of US dollars per 1 g). Relatively cheap ultra-dispersed systems (Al_2O_3, etc.) are preferable. To save fuel and minimize harmful emissions, water-fuel micro- and nanoemulsions are used, the permissible concentration of water in which depends on the size of the particles. (If the particle size does not exceed 1 μm, then stable operation of the heat engine is ensured with a water content of up to 80% [8].) Of particular interest in the creation of coatings are active minerals with shape and action memory [8]. Altair Nanotechnologies (USA) has developed nanoaccumulators—fast-charging (15 min) lithium-ion batteries with $Li_4Ti_5O_{12}$ electrodes. In hydrogen energy, carbon nanotubes are proposed for use in hydrogen storage devices.

Nanotechnology in nuclear power engineering

The main directions of introduction of nanotechnology in nuclear power engineering. The following directions are distinguished [8, 9] related to the creation and introduction of

- New structural and functional radiation-resistant heat-resistant materials capable of extending the service life of reactors (reactor vessels and internal steels): nanodispersed materials, zirconium alloys, ferritic-martensitic steels, dispersion-strengthened nanooxides, etc.;
- Highly efficient nuclear fuel, including high-density fuel with nanoadditives, mixed oxide fuel, development of the thorium-uranium fuel cycle;
- Technology to ensure comprehensive safety of nuclear power facilities and auxiliary equipment (including metal volumetric nanofilters for use in water treatment systems and coolant purification).

Nanotechnology (ultra-dispersed structures), as applied to the creation of new materials for nuclear technology, has been actively used since the early 1960s.

Structural materials. The greatest successes have been achieved in the field of creating new structural materials. For fast reactors (mainly of the BN type), specialists from Russia, the USA, Japan, China, France and Ukraine are considering radiation-resistant heat-resistant materials based on ferritic-martensitic steels dispersion-hardened with thermally stable nanooxides (3–5 nm), which have good strength and mechanical properties, and are capable of operating at neutron fluences (with kinetic energy greater than 0.1 MeV) up to 2×10^{16} cm^{-2} s^{-1} up to damaging doses of 160–180 dpa at temperatures of 370–710 °C (see [9] and references to this article). Pre-reactor tests of these steels in pilot-industrial conditions showed a multiple (up to 8 times) increase in heat resistance parameters compared to traditionally used steel [8]. The yttrium oxides Y_2O_3 used in such steels (for example, EP450 DUO) in the initial state are agglomerates several micrometers in size. Each agglomerate consists of crystallites (40–80 nm). As a result of extrusion, a polygonized structure with grains of about 3–5 μm is formed and the yttrium oxides inside the grains are crushed to 5–10 nm. The best creep resistance characteristics are achieved with the formation of the maximum number of uniformly distributed nanoclusters of 1–2 nm. Titanium and zirconium additions to steel contribute to the greatest degree to the refinement of yttrium oxide. Steels dispersion-hardened with yttrium nanooxides are characterized by the formation of a sufficiently thick oxide layer on the surface, which is associated with the ultrafine-grained structure of the steel [9]. (Also considered is the nanooxide Al_2O_3 [9].) With an increase in the amount of dispersed oxides (0.4% by weight of Y_2O_3 with characteristic dimensions of 40–100 nm and 0.06–0.30% by volume of Y_2O_3–TiO_2 with dimensions of 2.5–3.3 nm), the strength characteristics of steels are noticeably improved [8]. Steels, dispersion-strengthened with nanooxides, are used as a material for fuel element cladding in BN-600 and BN-800 reactors. The use of nanoborides (5–100 nm) in steel allows increasing the boron content by 3–4 times while maintaining the ductility and weld ability of steel [8].

Nanotechnology allows reversing the seemingly irreversible process of radiation embrittlement. This is achieved by forming a nanostructured sub lattice of short-range ordering clusters in a solid solution—vacancy traps and interstitiums with a period of 5–10 nm, commensurate with the mean free path of radiation point defects [8]. Irradiation of such alloys in a nuclear reactor leads to an increase in their strength while maintaining viscosity [8].

Fuel materials. In the task of creating highly efficient nuclear fuel (mainly for fast reactors), for which deep burn up is achievable, two subtasks are distinguished [8]:

- Creation of large-crystalline structures with a given porosity (oxide with nano-additives for BN-type reactors [9]), capable of retaining gaseous and volatile corrosive fission products, preventing transfer along grain boundaries to the fuel element cladding, in order to slow down the corrosion of the inner surface of the cladding.
- Activation of the sintering process of fuel pellets due to nano-additives. To achieve a coarse-crystalline state during pressing of UO_2, MOX (mixed oxide, $UO_{1.8}$–Pu_4O_7 or UO_2–PuO_2); MN (mixed nitride, UN–PuN) fuel ceramics, ultra-dispersed UO_2 powder (40 nm nanocrystallites) is added to it. As a result, the sintering temperature decreases by approximately 200 °C, the density of the ceramics increases and the grain size increases to 35–40 μm without deterioration of other characteristics. Fuel based on complex carbonitride UZrCN doped with nanodiamond is being considered for fast reactors [9].

3.2.2 Composite and Powder Materials

Composites in nuclear engineering. A composite material is a multi-component material made from two or more components with significantly different physical and/or chemical properties, which in combination lead to the appearance of a new material with characteristics that differ from the characteristics of individual components and are not their simple superposition. By varying the composition of the matrix and filler in the composite, their ratio, and the orientation of the filler, a wide range of materials with the required set of properties is obtained. The use of composites usually allows for a reduction in the weight of a structure while maintaining or improving its mechanical characteristics.

According to the matrix composition, polymer composite materials, composite materials with a metal matrix, and ceramic-based composite materials are distinguished [18].

Composites are widely used in equipment operating at elevated temperatures, pressures, intense radiation exposure, in chemically aggressive environments, in flows of corrosive and erosive liquids. Composites are used in aviation and space technology, deep-sea vehicles, transport, energy (including nuclear energy), in household goods, in medicine, and construction.

Powder materials in nuclear engineering

Fuel. There are two methods of manufacturing nuclear fuel. The most common is sintering fuel powders into tablets. Less common in power engineering is vibration compaction of fuel powders in a fuel element. Powders with a typical particle size of 50–300 μm are used.

Coatings. A number of materials have unique properties when microtechnology is used in their manufacture. In some practical cases, the use of microtechnology is quite sufficient and justified, although nanotechnology has a clear advantage. This applies, for example, to corrosion- and erosion-resistant tungsten coatings (40–100 μm thick) of the inner and outer shells of fuel elements of lead-cooled reactors [19].

At the dawn of nuclear technology development, the possibility of using coatings of vibrationally compacted spherical MOX fuel particles (a "snowball-like" coating) was considered [20]. Tungsten, tantalum, molybdenum, niobium, titanium, etc. were proposed as materials for such coatings [20]. Given certain difficulties in the production of vibrationally compacted fuel containing a free oxygen getter, associated with ensuring uniform mixing of MOX particles and the getter, it is possible to propose using coatings of MOX particles with metallic beryllium [21].

By 2018, specialists from the Keldysh Center, Moscow Aviation Institute, and the A.A. Baikov Institute of Metallurgy and Materials Science of the Russian Academy of Sciences (Moscow, Russia) had developed cladding for WWER-type reactor fuel pins with a hafnium dioxide or zirconium dioxide coating, as well as multilayer (three-layer) coatings based on nanomaterials. The layers are Ni–Co–Cr–Al–Y alloy, ZrO_2–7% Y_2O_3, HfO_2–9% Y_2O_3 compounds, etc. [22]. The coating thickness does not exceed 120 μm [22].

Specialists from the National University of Science and Technology MISiS have developed corrosion-resistant and radiation-resistant three-layer shells for fast reactors with lead coolant [23]. The shell contains a central layer of vanadium alloy 90.3% (weight) V–4.9% Ti–4.8% Cr [24]. This layer is surrounded by layers of corrosion-resistant steel Kh17N2 with low nickel content [23].

Previously, the author of this article proposed tungsten coatings for fuel pin cladding made of EP-823 steel (Kh12MVFBR) with an increased silicon content and a low nickel content for fast reactors with lead coolant [19].

Technology for manufacturing finely dispersed UO_2–U fuel. Scientists from the V.G. Khlopin Radium Institute have proposed a technology for manufacturing tablet-shaped dense and heat-conducting fuel based on UO_2 and metallic uranium with strong adhesion of particles of different phases to increase the content of fissile material in fuel [25]. Crushed (50–30 μm) UO_2 is mixed with nanodispersed (80 nm) hydride of enriched uranium. The amount of hydride in terms of metallic uranium is 5–50%. The mixture is then homogenized by grinding and mixing together. The resulting mixture is heated (1 °C/min) to 300–330 °C in a vacuum (10^{-4}–10^{-5} mmHg). This decomposes uranium hydride into metal. The next step is pressing the tablets using an organic binder, sintering them in a vacuum (10^{-4} mmHg) at 1500–1550 °C and cooling in a furnace. After cooling in the furnace, the fuel pellets remain stable in air. At the same time, the tetravalent state of uranium in the oxide composition is preserved. U particles are uniformly distributed in the volume of the UO_2 matrix with the formation of continuous structural networks, which helps to increase the strength of such tablets [25]. When mixing fine-grained uranium dioxide (82%) with nanosized uranium hydride (18%), the density of the calcined tablets was 11.9 g/cm^3, and the thermal conductivity was 12 W/(m K) (under normal conditions) [25].

Vibration-compacted oxide fuel. Since 1980, specialists from the A.A. Bochvar All-Russian Research Institute of Inorganic Materials and the State Research Center of Atomic Reactors have been considering the possibility of using vibration-compacted MOX fuel with additives of 3–10% (optimum 7%) of finely dispersed metallic uranium powder (100 μm) as a getter of free oxygen for BN-type reactors [26, 27]. In the fuel pins of experimental fuel assemblies of the BN-600 reactor, a burn up of 9.6% was achieved, and in individual fuel pins, 10.8% [26]. Uneven mixing of uranium and MOX during mass production of fuel did not allow the elimination of intense corrosion of the inner surface of the BN-600 reactor fuel pin cladding.

Since it is difficult to control the uniformity of mixing of MOX powders and metallic uranium in a relatively large volume of fuel pin (with vibration compaction), it makes sense to move to small volumes (pellet fuel).

MN nanopowder-based pellet fuel. Further improvement of fuel characteristics while maintaining a high melting point is possible by switching to MN nanopowder-based fuel with a particle size of 100–500 nm. This allows reducing fuel porosity from 25 to 5%, which means increasing reactor power by 20%.

To achieve a density of 85–95% of the theoretical value, the required sintering temperature is 1800–1900 °C and the time is 10–30 h. In this case, up to 15% of the plutonium evaporates from the fuel [28]. Electromagnetic compaction methods make it possible to reduce the sintering time and temperature [24]. So far, only the possibility of obtaining ultra-dispersed UN powder has been experimentally substantiated [28].

3.2.3 Coolant

It is proposed to use liquid lead of natural isotopic composition as a coolant for BREST reactors: 1.4% ^{204}Pb–23.6% ^{206}Pb–22.6% ^{207}Pb–52.4% ^{208}Pb. This is the average isotopic composition for all known lead deposits.

To reduce the void effect of reactivity in high-power reactors, it will be necessary to significantly increase the proportion of the isotope ^{208}Pb in the coolant. This is possible by using lead extracted from thorium ores.

For example, lead extracted from monazite of pegmatites of thorium ore granites has the composition: 0.01 to 0.076% ^{204}Pb–0.89 to 26.43% ^{206}Pb–0.35 to 4.11% ^{207}Pb–69.15 to 97.74% ^{208}Pb. The concentration of the ^{207}Pb isotope is minimal, and that of ^{208}Pb is maximal. This is ideal for minimizing the void effect of reactivity in fast reactors of medium and high power. Figure 3.2 shows the dependence of the content of the C_{208} isotope ^{208}Pb in lead on the content of C_{Pb} lead in monazite of the Ukrainian Shield (according to [29]).

Note that when introducing reactivity $\rho > \beta$ (β is the effective fraction of delayed neutrons), which corresponds to reactor acceleration on prompt neutrons, the change in reactor power or neutron density n with time t can be estimated using the point kinetics equation

$$dn/dt \approx (\rho - \beta)nk/L,$$

where k is the effective neutron multiplication factor; L is the average lifetime of prompt neutrons, $L = L_\infty P$; L_∞ is the average lifetime of prompt neutrons in an infinite medium; P is the probability of no leakage.

The average lifetime of prompt neutrons in an infinite medium is $L_\infty = v^{-1}\,\Sigma_a{}^{-1}$. Here v is the average neutron velocity; Σ_a is the macroscopic neutron absorption cross-section.

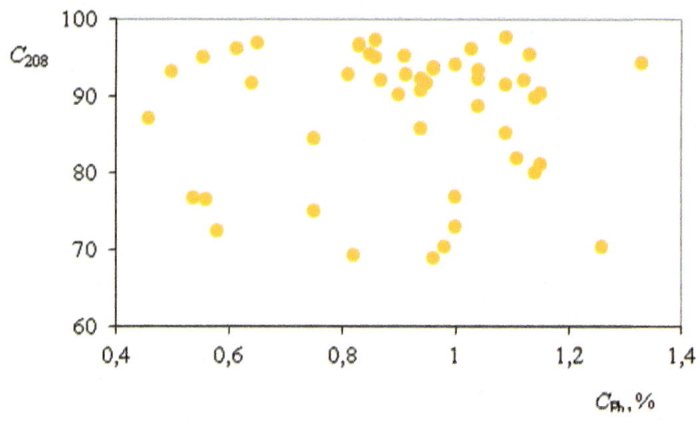

Fig. 3.2 Dependence of the content of the isotope ^{208}Pb in lead on the content of lead in monazite (according to data from [29])

Using lead-208, which weakly absorbs neutrons, leads to an increase in L and, as a consequence, to a decrease in *dn*/*dt* for given ρ and β. This is important for increasing safety in emergency modes with the introduction of positive reactivity.

3.3 Methods

3.3.1 Research Object

Some projects of fast reactors with liquid metal coolant satisfy the requirements for energy systems of the future. Lead coolant reactors with mononitride fuel are the most preferable. The reactor core is considered as a system of multiple elements that are interconnected with each other. The study of such systems is carried out within the framework of system analysis and is solved based on the methods of operations research related to the construction, development and application of mathematical models of decision making. Elements of operations research (mathematical programming, multicriteria optimization, Markov decision-making models, decision-making procedures under risk and uncertainty, game theory, etc.) are used to solve the problem of selecting the optimal physical characteristics of the core [21].

The core is contains fuel assemblies, the geometry of which is similar to that used in the BREST-OD-300 reactor under construction [7]. The core contains three profiling zones, differing in the diameter of the fuel elements used [7]. The height of the fuel column of the core is 1.1 m. The outer radii of the profiling zones are 2.23; 2.90 and 3.22 m, respectively.

The side reflector surrounding the core contains tubes with radioactive waste to be transmuted. The first rows of reflector assemblies, located closer to the core, contain technetium-99, the last rows of reflector assemblies contain technetium-99 and carbon-14. Moderation of neutrons on carbon nuclei helps to increase the efficiency of transmutation of both technetium-99 and carbon-14.

3.3.2 Modeling of Emergency Modes

When designing new generation reactors, the most dangerous emergency situations accompanied by failure of emergency protection (ATWS) should be considered. They can be combined into a small number of groups. The classification feature of the combination is the disturbance initiating the emergency mode. For fast reactors with liquid metal cooling—the disturbance of the flow rate—δG, the coolant temperature at the core inlet—δT, the reactivity—δρ. An emergency situation with the drying of the core or its part—LOCA WS (loss of coolant accident without scram) is singled out into a separate group:

- The disturbance $\delta\rho > 0$ corresponds to the emergency mode group TOP WS (transient overpower without scram). This mode was initiated by the injection of positive reactivity.
- The disturbance $\delta G > 0$ corresponds to the emergency mode group OVC WS (overcooling accident without scram). This mode was initiated by an increase in the coolant flow.
- The disturbance $\delta G < 0$ corresponds to the emergency mode group LOF WS (loss of flow without scram). This mode was initiated by a decrease in the heat carrier flow in the primary circuit (for example, when the main circulation pumps are de-energized).
- The disturbance $\delta T > 0$ corresponds to the group of emergency modes LOHS WS (loss of heat sink without scram). This mode was initiated by an increase in temperature at the inlet to the core (for example, when the secondary circulation pumps are de-energized).
- The disturbance $\delta T < 0$ corresponds to the group of emergency modes OVC WS (overcooling accident without scram). This mode was initiated by a decrease in the coolant temperature at the inlet to the core.

The uncertainty of the scenarios for the development of emergency modes is determined by the uncertainties of the disturbance initiating them, the law of change of disturbance over time, the delay time of a particular disturbance (when events overlap).

3.3.3 Software

The following programs and software packages developed by the author are used as software: Dragon-M (for solving optimization problems according to a selected criterion with specified constraints, including safety functionals, in a deterministic formulation or under uncertainty, including scenarios for the development of emergency situations); FRISS (for modeling emergency situations of the ATWS type and their combinations in the point kinetics approximation taking into account feedback on average temperatures) [21]; codes implementing various methods for solving discrete multi-criteria problems; the well-known MCU code [30] for precision neutron-physical calculation, hierarchical models with adjustable accuracy developed by the author based on it, the well-known WIMS code for neutron-physical calculation of an infinite reactor lattice [31].

3.4 Results

3.4.1 The Problem of Selecting Preferred and Acceptable Materials

Modeling the structure and properties of nuclear engineering materials involves optimizing the fuel properties and creating new-generation materials with specified properties. In order to solve this problem, the author carried out another upgrade of the Dragon-M calculation and optimization software code [21]. The properties of the core materials were considered as control parameters of the optimization problem. Among them are the density, heat capacity, and thermal conductivity of the fuel; thermal conductivity and neutron absorption cross-section by the fuel element cladding material; neutron interaction cross-sections (absorption, scattering, moderating ability) with coolant nuclei. Optimization problems for the minimum void effect of reactivity under the most dangerous scenario of its implementation were considered; as well as problems without a criterion for finding the area of permissible layouts of the active zone of a fast reactor. Using the Dragon-M code, optimal and permissible values of these parameters were obtained. Based on the obtained parameter values, real materials were selected.

3.4.2 Fuel

Range of acceptable fuel property values. The results of searching for acceptable physical properties of fuel (density and thermal conductivity) are shown in Fig. 3.3. The range of acceptable values (area) is shaded. Inside this area there is a point corresponding to U–Pu fuel. The melting point of such fuel is low. Alloying metallic fuel with zirconium does not solve the problem. Outside the area there is a point corresponding to MN fuel. Cermets fuel UN–PuN–U is near the boundaries of the acceptable area.

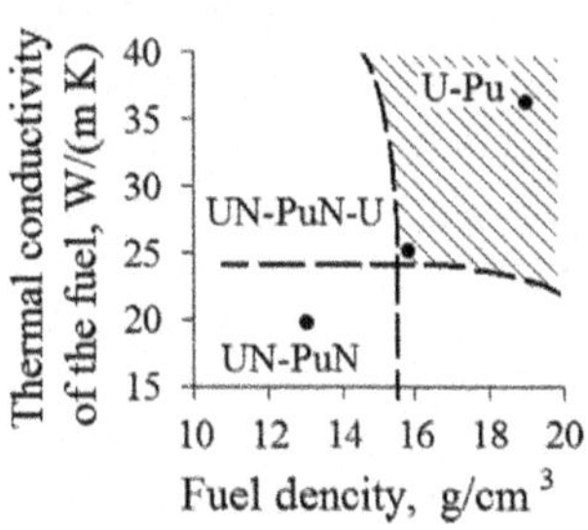

Fig. 3.3 The range of valid values of nuclear fuel (according to article [32, 33])

Thus, the obtained properties correspond to the cermets fuel UN–PuN–U. However, the melting point of such fuel is significantly lower than the melting point of mononitride UN–PuN. Such fuel is unacceptable.

Mathematical modeling of emergency modes has shown that to ensure the same role of the Doppler coefficient of reactivity in different emergency situations, the density and thermal conductivity of the fuel must be higher than for mononitride. At the same time, density and thermal conductivity values significantly lower than for metallic fuel are acceptable. The transition to "classic" cermets fuel will optimize the density and thermal conductivity; ensure the same role of the Doppler coefficient in different emergency situations, but leads to an unacceptable decrease in the maximum permissible fuel temperature.

How to resolve the conflict in obtaining the required (specified) fuel properties when using cermets?

It is proposed to use pellet fuel based on micrograins of mixed mononitride fuel and metallic uranium nanopowder. The nanopowder fills the pores between the micrograins. Its content is up to 20% by volume or up to 39% by weight.

When using such fuel, the role of the Doppler reactivity coefficient in different emergency modes is the same. This opens up additional opportunities for optimizing the core layout. When using different types mixed oxide—MOX, mixed mononitride—MN, mixed metal—U–Pu, mixed mononitride fuel with additives of metallic uranium nanopowder, the maximum fuel temperature T_f during operation of the BREST reactor at nominal power:

$$T_f(MOX) > T_f(MN) > T_f(MN - U) > T_f(U - Pu).$$

Maximum fuel temperature in steady-state mode of natural circulation of the coolant:

$$T_f(MOX) < T_f(MN) < T_f(MN - U) < T_f(U - Pu).$$

Fuel based on ceramics and metallic uranium nanopowder. For this reason, pellet fuel based on mononitride micrograins (ceramics) and metallic uranium nanopowder can be considered. Metallic uranium nanopowder fills the pores between the mononitride micrograins [32]. The mass fraction of nanopowder is 39% [32]. The characteristic size of micrograins is from 50 to 300 μm. The characteristic size of nanopowder particles is from 40 to 100 nm. By varying the size of the ceramic micrograins, it is possible to change the ratio of ceramics and metal in the fuel. One of the most preferable options corresponds to a volume fraction of metallic uranium nanopowder equal to 20% (which corresponds to a mass fraction of 39%) [32].

Nanopowder of metallic uranium is a good getter of free nitrogen released from the fuel during reactor operation. The presence of free nitrogen increases the rate of corrosion of the inner surface of the shells under the action of some fission products migrating to the shell.

3.4.3 *Absorber Pellets in the Fuel Element. Analysis of a Hypothetical Emergency Situation with the Destruction of Fuel Element Cladding*

Probabilistic, deterministic and phenomenological approaches are used in the justification of the safety of nuclear reactors [34]. The deterministic approach can be formally defined as a phenomenological plus "invented accidents". The term "invented accidents" was proposed by the developers of the BREST concept for the justification of safety [35]. As will be shown, in the BREST type reactors improved by the author (using new fuel, coolant, fuel element cladding) in all emergency modes that cannot be excluded deterministically, a large temperature margin is maintained before lead boils, mononitride decomposition in the cermets fuel and damage to the fuel element cladding. Maintaining the integrity of the cladding is considered the most vulnerable link in ensuring the self-protection of lead-cooled reactors.

In the upper part of the fuel elements of the BREST-OD-300 reactor, above the fuel column, there is an absorber tablet 25–35 mm high made of tungsten boride [7].

Let us consider a hypothetical case, assuming that the fuel pin cladding of the BREST-OD-300 reactor with mononitride fuel has been destroyed. We will assume that the destruction of the fuel pin cladding was initiated by the superposition of the LOF WS process (the greatest increase in cladding temperature) with some other process from the ATWS. In this case, the lead flow through the core is small: approximately at the level of natural circulation. Let us analyze two scenarios for the development of an emergency situation. The first (hypothetical): the pellets are sintered into a single rod and do not crumble in the absence of cladding.

Although the relatively low operating temperature of the mononitride does not promote pellet sintering, the low velocity of lead during natural circulation helps to preserve the fuel column in the absence of cladding. The density of structural steel is less than that of lead, therefore, the material of the cladding and spacer grids will float even in the absence of coolant circulation. A conservative estimate of the reactivity effect with complete removal of the cladding material from the active zone of an infinite-radius reactor yields a value of 9.8 β (removal of 72.8% of steel), cladding and spacer grids (81.7% of steel)—plus 11.0 β.

Another scenario (realistic): the fuel column disintegrates into pellets. Depending on the lead velocity, density and height of the fuel pellets, they can float or sink in the core (Fig. 3.4).

The same remark applies to the tungsten carbide absorber pellets located in the upper part of the BREST-OD-300 reactor fuel elements [7, 36]. Positions "*a*" and "*b*" in Fig. 3.4 are the dependence of the maximum height h of a pellet on the density γ, which guarantees the pellet's floating in the coolant flow: the coolant velocity is indicated next to the curve, the area under the curve—floating; position "*c*" in Fig. 3.4 is the dependence of the minimum velocity v of the coolant, which guarantees the floating of the tablets, on their height h: the dotted line corresponds to the absorber tablets made of WC, the solid lines—to fuel tablets of different density, the density is indicated next to the curve (g/cm^3), the area above the curve—floating.

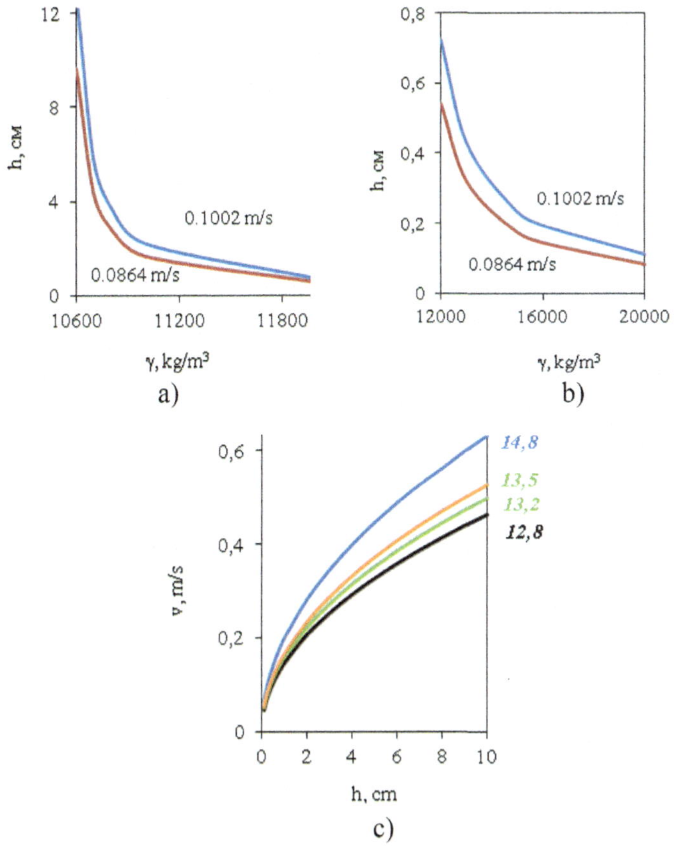

Fig. 3.4 Possibility of a WC-pellet floating in the coolant flow

If the fuel column disintegrates into pellets, the coolant flow can carry out fuel pellets no higher than 0.3 cm from the core, i.e. in BREST-OD-300 reactors, the lead velocity in the natural circulation mode is clearly insufficient for the coolant flow to carry out fuel fragments from the core. The reactor acceleration on prompt neutrons when the cladding material is carried out from the core is prevented by the fall of the absorber pellets into the core. At low lead velocity (natural circulation); the WC pellets sink in the core. The absorber pellet can get stuck and not get into the inter-fuel space. Therefore, it is better to use the absorber not as a single pellet, but to assemble it from 3–4 pellets.

3.4.4 Coolant

The use of a lead coolant based on the isotope ^{208}Pb opens up wide possibilities for improving the neutron balance and increasing the safety of fast reactors. The nuclei of ^{208}Pb are doubly magical. They are distinguished by increased stability and, therefore, lower activation in the active zone. Inelastic processes on ^{208}Pb nuclei are realized with a lower probability than on other stable isotopes of lead, i.e. ^{208}Pb absorbs neutrons worse and slows down (due to inelastic scattering) worse.

The worst neutron moderation (on ^{208}Pb nuclei) will allow for the intensification of threshold fission of heavy nuclides with an even number of neutrons (^{237}Np, ^{238}U, ^{241}Am, ^{243}Am, etc.) by increasing the neutron energy in the reactor core and minimizing the void and density effects of reactivity when using a coolant based on the ^{208}Pb isotope.

Enrichment of lead with the isotope ^{208}Pb is an expensive operation, which prevents the large-scale use of such a coolant in power engineering. Depending on its origin, lead can have a different isotopic composition. For cooling BREST-type reactors, it is proposed to use lead, which is a mixture of four stable isotopes: 1.4% ^{204}Pb–23.6% ^{206}Pb–22.6% ^{207}Pb–52.4% ^{208}Pb.

The isotope ^{204}Pb is of non-radiological origin, the isotopes ^{206}Pb, ^{207}Pb and ^{208}Pb are of radiological origin: they are the last (stable) nuclides of the natural radioactive series of uranium-radium, uranium-actinium and thorium, respectively. By now, approximately 1/5 of ^{232}Th has decayed into ^{208}Pb. Thus, the possibility of isolating lead with a high content of the isotope ^{208}Pb during the extraction of thorium ores is attractive.

Natural uranium consists mainly of two isotopes: ^{235}U (0.72%) and ^{238}U (99.28%)—the progenitors of radioactive series. Therefore, uranium ores contain lead with a high concentration of the isotope ^{206}Pb and a much lower concentration of ^{207}Pb. In nature, thorium occurs in the form of minerals together with uranium, therefore, the by-product of thorium ores will be a mixture of the isotopes ^{206}Pb, ^{207}Pb and ^{208}Pb with a high concentration of the latter (up to 98% [29]).

Since the isotopic composition of lead contained in thorium and uranium ores can vary widely, the problem of selecting preferred deposits is relevant.

The main ore from which lead is mined is galena (lead glitter). It is a component of polymetallic (lead–zinc) ores with lead and zinc content of several to 10% or more. The content of the ^{208}Pb isotope in the lead of these ores is 52.4%. Thus, the attractiveness of thorium ores for obtaining lead with high ^{208}Pb content is obvious, the economic benefit is enormous. For example, rich polymetallic ores contain up to 5% lead with a ^{208}Pb content of 52.4%. In other words, polymetallic ores contain up to 2.5% ^{208}Pb, the extraction of which will require isotopic separation. Even minerals that are poor in thorium (and therefore in the accompanying lead), such as monazites, contain about 1% lead, which consists almost entirely of the isotope ^{208}Pb, the isolation of which practically does not require isotopic separation. Obviously, the most preferable is lead with the maximum concentration of the isotope ^{208}Pb and the minimum concentration of the isotope ^{207}Pb. (The number of neutrons in the ^{207}Pb nucleus is 125, i.e., it precedes the magic number corresponding to a completely filled neutron shell. As a result, ^{207}Pb, among all other stable lead isotopes, has the maximum cross-section for neutron attachment, i.e., the maximum cross-section for inelastic processes—absorption and inelastic scattering. Therefore, when using a lead-based coolant with a high content of ^{207}Pb, one should expect the maximum void effect of reactivity.) Lead isotopes ^{204}Pb and ^{206}Pb with an even number of neutrons, due to the pairing effect, have smaller cross-sections for inelastic processes than ^{207}Pb, and larger ones than the doubly magic nuclei ^{208}Pb.

Thus, there is no need for radical minimization of the ^{204}Pb and ^{206}Pb isotope content in the coolant. Their minimization makes sense if it is implemented by increasing the share of the doubly magic isotope ^{208}Pb in lead. If the decrease in the share of ^{204}Pb and ^{206}Pb occurs due to an increase in the concentration of ^{207}Pb, the content of ^{204}Pb and ^{206}Pb isotopes in lead, on the contrary, should be maximized. Thus, the target functionals can be considered with weighting factors determined subjectively (by the decision maker). In this case, the weighting factors reflecting the significance of the criteria for ^{204}Pb and ^{206}Pb concentrations will be less than for ^{207}Pb and ^{208}Pb concentrations.

The most preferred deposit of the Ukrainian Shield has the following characteristics: lead content in the ore is 1.09%, lead isotopic composition is 0.013% ^{204}Pb–0.89% ^{206}Pb–0.35% ^{207}Pb–97.74% ^{208}Pb. Thorium content in the ore is 11.26%, uranium—is 0.068% [29].

The results of the calculation of the void and density effects of reactivity obtained for a BREST reactor with an infinite radius of the core are presented in Fig. 3.5. It is assumed that the bubbles in the core rise from the bottom to the top.

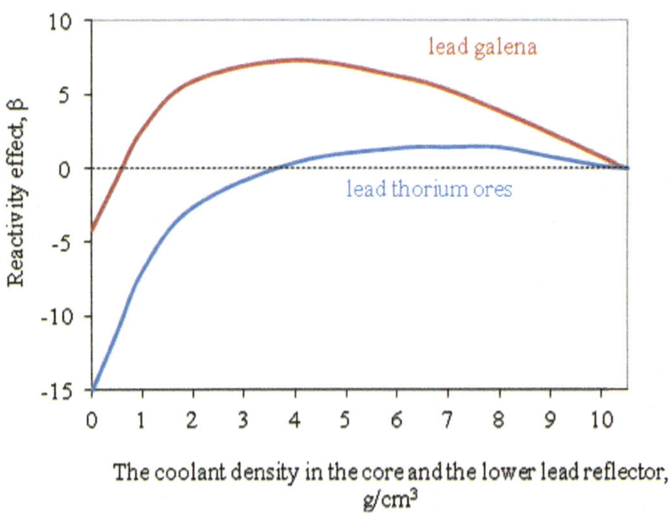

Fig. 3.5 The reactivity effect in case of drainage "from below up" (involvement of bubbles in the core) for BREST of big capacity (calculations for MCU)

Lead extracted from thorium ores and lead extracted from galena is considered as the coolant.

Figure 3.5 shows the dependence of reactivity effects on the density of lead in the core. Zero density of lead corresponds to the void effect of reactivity.

The void effect of reactivity is negative when draining the core when using lead galena and lead thorium ores. When using lead galena, the maximum density effect of reactivity is slightly more than 7 β with a decrease in the coolant density to 4 g/cm^3. When using lead thorium ores, the maximum density effect of reactivity is slightly more than 1 β with a decrease in the coolant density to 6.5–8.0 g/cm^3.

The transition to lead with a high content of the isotope ^{208}Pb leads to a noticeable reduction in the void and density effects of reactivity (Fig. 3.5). However, to reduce these effects to a safe (negative) value, additional measures must be taken. Such measures may include the use of tungsten coatings on fuel pin cladding or the placement of an additional neutron absorber (B_4C or WB) in the central part of the core.

3.4.5 Fuel Element Cladding Materials

Assessment of the reactivity effect during the removal of corrosion and erosion products of structural steels from the BREST reactor core. In reactor technology, the corrosion and erosion effect of the coolant on structural materials is one of the key problems, the solution of which determines the possibility of using a particular technology.

For the BREST-OD-300 reactor, the problem of corrosion and erosion of structural materials has been solved [7, 36].

The problem of corrosion in BREST-type reactors has been solved by using corrosion-resistant silicon steels and optimal oxygen content, and erosion by low coolant velocity [7, 36].

The core of the BREST-OD-300 reactor contains about 35.5 tons of steel, about 4.3 g of corrosion products per hour or 31 kg of structural steel per year are removed from the core [36], which corresponds to a reactivity input of 0.0118 β. It is important to note that with such a corrosion rate, self-protection against severe accidents is easily ensured throughout the entire campaign, and the standard coolant purification system should easily cope with such a quantity of slag [36].

The increase in the capacity of power reactors occurs due to the increase in the volume of the core, mainly due to the increase in the diameter of the core at a constant height. This is due to the desire to limit the void effect of reactivity. Therefore, for the BREST-1200 reactor, the rate of removal of corrosion products can be estimated at 124 kg of structural steel per year. (This is only from the core.) Estimates of economic damage should be made not in the cost of the removed steel, but in the cost of damaged fuel assemblies containing at least one damaged fuel element.

The processes of corrosion and erosion are always associated with troubles and problems. If these processes are inevitable, can they be used for good?

Thermal neutron power reactors typically use a burnable absorber. This is a neutron absorber that is consumed during reactor operation. This partially compensates for the loss of reactivity due to the burnout of nuclear fuel. Excess reactivity is necessary to ensure the required fuel burnout in the reactor core. These are stationary absorbers that are unloaded from the core along with the fuel during refueling.

Burnable absorbers provide reactivity control for different fuel cycles, long-lived radionuclide disposal, and reactor safety. When selecting the ideal burnable absorber type and design, the resulting impact on reactor fuel cycle design and cost, reactivity, thermal hydraulics, fabrication, and radiation response must be considered [37]. Widely used burnable absorbers are gadolinium (Gd_2O_3) and erbium (Er_2O_3) oxides. Gadolinium oxide is used in WWER reactors, and erbium oxide is used in RBMK reactors.

In addition, a boron (B_4C) burnable absorber is usually used in the composition of the coolant of a thermal neutron reactor. B_4C can be either a natural mixture of boron isotopes containing about 19% of the isotope ^{10}B and 81% of ^{11}B, or a mixture enriched in the isotope ^{10}B with a content of this isotope up to 75%. In fast reactors, burnable absorbers are usually not used. However, a "corrosive neutron absorber" can be proposed for fast reactors.

Minimizing the rate of corrosion and erosion of structural steels is important both from an economic point of view (to a certain extent it determines the service life of fuel elements) and from a safety point of view (firstly, it can lead to the destruction of fuel element cladding in emergency situations, and secondly, it is characterized by a positive reactivity effect when removing corrosion and erosion products from the active zone and depositing them in special filters or in cold sections of the circulation circuit).

So, is it possible to turn the natural processes of corrosion and erosion into a positive action? Any non-fissile material is a parasitic neutron absorber. Therefore, the removal of corrosion and erosion products from the reactor core is accompanied by the introduction of positive reactivity. This effect can partially compensate for the decrease in reactivity due to fuel burnout. Thus, from the point of view of neutron-physical processes, corrosion-erosion mass transfer, for example, in BREST reactors can play the same role as a burnable absorber in thermal reactors. (In BREST reactors, fuel burnout is compensated by its production: the breeding ratio in the core is ≈ 1).

Figure 3.6 shows the change in reactivity during the removal of corrosion and erosion products from the active zone of a BREST reactor, characterizing the efficiency of the "corroding neutron absorber". Here and below, β is the effective fraction of delayed neutrons. The mononitride UN–PuN was used as fuel, lead of natural isotopic composition was used as a coolant, and EP-823 steel was used as the material for the fuel pin cladding [7, 36]. As follows from Fig. 3.6, the positive effect of reactivity due to corrosion and erosion of steels is insufficient to compensate for fuel burn up. The assessment was made assuming a constant rate of removal of corrosion and erosion products and a conservative assumption of their complete removal from the core.

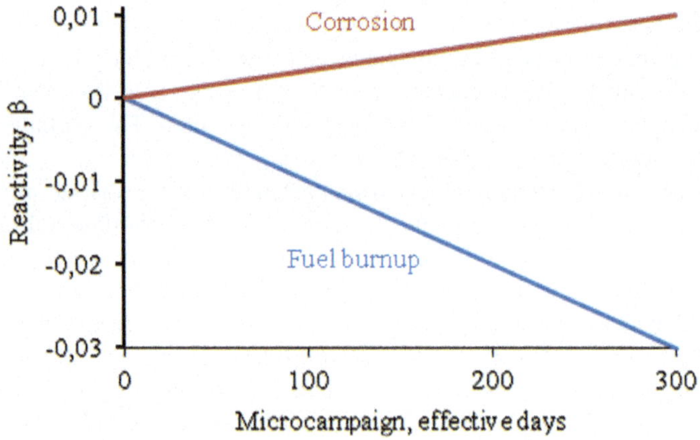

Fig. 3.6 Change in reactivity over a micro campaign as a result of the removal of corrosion and erosion products from the core and fuel burnout

Thus, with a change in reactivity over a micro campaign of about 0.03 β, the corrosion and erosion process can compensate for the decrease in reactivity due to fuel burnout by approximately 40%.

Thus, structural steel can function as a "corroding neutron absorber" that is consumed during reactor operation. As the neutron-absorbing steel corrodes and erodes, reactivity increases. As a result, the loss of reactivity due to nuclear fuel burnout is partially compensated (by approximately a third).

Chemical reaction of water vapor with chromium. BREST reactors have a two-loop energy conversion scheme [7, 36]. The first loop contains liquid lead. The second loop contains superheated water vapor.

In all severe accidents of light-water thermal reactors, a zirconium-steam reaction was realized. This is an exothermic chemical reaction between zirconium and water vapor, which occurs at high temperatures under conditions of contact between zirconium structural elements and water [37]. Water vapor at high temperatures reacts with zirconium:

$$Zr + 2H_2O \rightarrow ZrO_2 + 2H_2.$$

This produces hydrogen at a rate of about 0.491 L/g of zirconium reacted and releases heat of 6530 kJ/kg. Hydrogen can explode, making the accident worse. Mixtures of hydrogen with oxygen and air are explosive and are called detonating gas. Detonating gas spontaneously ignites at atmospheric pressure at a temperature of 510 °C.

The zirconium vapor reaction begins at approximately 861 °C and at 1200 °C it begins to develop very rapidly, as the released heat further heats the zirconium and the reaction becomes self-sustaining [37].

In addition to the release of hydrogen and heat, the reaction is accompanied by a loss of strength of the fuel element cladding and a decrease in their initial thickness due to the rapid oxidation of zirconium. In 10–12 min after the start of the self-sustaining steam-zirconium reaction, the fuel element cladding oxidizes to a thickness of 0.10–0.15 mm and begins to melt.

Metals of medium activity, located between hydrogen and aluminum in the electrochemical series of metal activity (the series of standard electrode potentials): titanium, manganese, zinc, chromium, iron, nickel, lead, etc., heated to 600–800 °C, react with superheated water vapor to form hydrogen. These metals are often used as an alloying additive and structural steel.

Chromium is part of the structural steel of the BREST reactor (16% by weight). When superheated water vapor interacts with chromium, a chemical reaction similar to the zirconium-steam reaction is possible:

$$2Cr + 3H_2O \rightarrow Cr_2O_3 + 2H_2.$$

Elemental chromium practically does not react with water at low temperatures. The reaction occurs at a temperature of not less than 600–700 °C. Almost all metals used as alloying additives to EP-823 steel also react at high temperatures with superheated water vapor to form hydrogen.

To prevent such reactions, it is necessary to isolate the steel from steam. In BREST reactors, certain oxygen content is maintained in the lead coolant. This promotes the formation of protective oxide films on the surface of the fuel elements. The alloying additive, silicon, is a good deoxidizer. It promotes the natural formation of such protective films during reactor operation. At low oxygen concentrations in lead, films do not form. As the proportion of oxygen in lead increases, these protective films peel off.

Tungsten does not react chemically to form hydrogen. It is attractive as a protective coating for steel. Tungsten is immediately after hydrogen in the stress series. Under normal conditions, tungsten is chemically stable. It does not react with water and has high corrosion resistance.

Like silicon, tungsten promotes the formation of a protective oxide film on the surface of the cladding (to a lesser extent, since it retains oxygen slightly worse: the bond energy of $SiO = Si + O$ is 8.24 ± 0.10 eV, $WO = W + O$ is 6.94 ± 0.43 eV [38]). Due to the leaky nature of the circuit, oxygen will always be present in the lead. Moreover, as the temperature increases (to values characteristic of BREST reactors), in the presence of oxygen, tungsten oxidizes first to WO_2, then to W_4O_{11} and WO_3. This oxide film holds well (at least up to a temperature of 923 °C, significantly exceeding the operating temperature for BREST reactors) and serves as an additional naturally formed protective coating.

However, there is practically no oxygen in the active zone of a fast reactor, which means that the released free hydrogen does not pose a danger, and a hydrogen explosion is impossible. The main problem is associated with additional heat release in chemical reactions with the release of hydrogen and with the fact that hydrogen in the active zone slows down neutrons.

In the event of a hypothetical reactor or heat exchanger failure, free hydrogen may be released into the atmosphere. This will lead to an explosion and an aggravation of the accident.

Another possibility for increasing the corrosion resistance of the inner surface of fuel pin cladding involves self-organization of the coatings of the inner surface of fuel pin cladding. Specialists from the Leipunsky Institute of Physics and Power Engineering and the Dollezhal Research Institute of Power Engineering have analyzed the possibility of increasing the corrosion resistance of the inner surface of fuel pin cladding by the natural formation of protective coatings during the chemical interaction of the cladding material with the liquid metal sub layer [39, 40]. It is proposed to use liquid lead saturated with zirconium or a liquid eutectic alloy of lead with magnesium, also saturated with zirconium, as the material for the sub layer between the fuel and the cladding (the most preferable eutectic is of the following composition, % (mass): 97.53 Pb–2.25 Mg–0.2 Zr). During the operation of the fuel element, a protective coating based on zirconium nitride spontaneously forms on the inner surface of the cladding. In addition, "self-healing" of accidental damage to the coating is ensured [39, 40].

Tungsten coatings of fuel pin cladding. When using tungsten coatings of fuel pin cladding, the rate of corrosion and erosion should be significantly lower. Tungsten intensively absorbs fast neutrons. However, when using lead coolant with a high concentration of lead-208 and cermets fuel, the neutron balance of the core is so good that it allows the use of fuel pin cladding even entirely made of tungsten.

The dependence of the neutron multiplication factor k_{inf} in an infinite medium on the mass fraction w of tungsten in the fuel element cladding material is shown in Fig. 3.7. As an example, the fuel element lattice of the central zone of the BREST-OD-300 reactor is considered [7]. Two variants of the lead coolant used are considered: a coolant based on lead from thorium ores (80% ^{208}Pb) and a coolant based on lead from galena.

The presence of such an additional absorber (tungsten shells or coatings) reduces the void effect of reactivity. To reduce the rate of corrosion of the inner surface of the shells, steel with a tungsten coating on the inside can be used as a shell material. The deterioration of the neutron balance, if necessary, is easily compensated for by increasing the proportion of ^{208}Pb in the coolant (for example, when using lead extracted from thorium ores). There is a technology for low-temperature plasma spraying of one or more layers of tungsten with subsequent heating (up to 150 °C), which increases the reliability of the coating. When spraying tungsten on the inner surface of the shell, even its peeling will not lead to a noticeable introduction of positive reactivity.

Tungsten coatings are applied by low-temperature plasma spraying. This technology is well-established. Tungsten is applied to a steel surface of any geometric shape without significant heating of the steel coating (no more than 100–150 °C at a plasma jet temperature of several thousand degrees) [19]. It is possible to obtain mechanically stable multilayer coatings (with a minimum tungsten layer thickness of 40–100 μm), combining different protective properties, not peeling off during

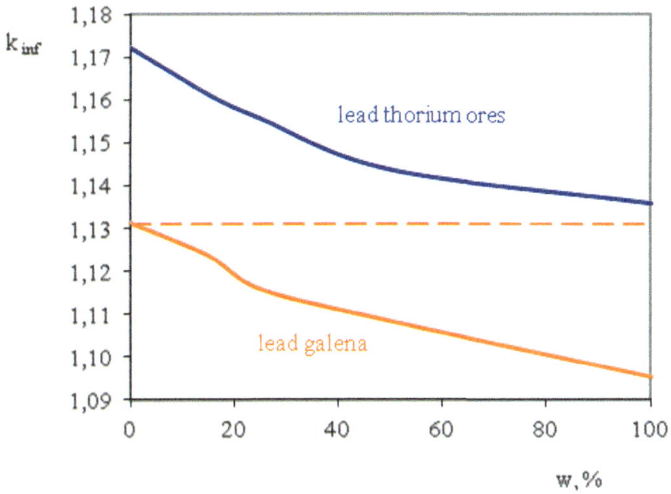

Fig. 3.7 Dependence k_{inf} (w)

bending and thermal expansion—compression of the main surface [19]. Ensuring high adhesion density (with tungsten spraying—more than 80 MPa) with low porosity of the coating (less than 1%) [19].

Tungsten is brittle under normal conditions. However, at temperatures exceeding 200–400 °C, it becomes plastic [41].

The high chemical activity of tungsten is explained by the structural features of the electron shell. The activity of tungsten is manifested in the fact that it enters into reactions with the overwhelming majority of elements, forming many simple and complex compounds. Tungsten forms alloys and stable intermetallic compounds with almost any metal [41]. It practically does not interact with nitride fuel at temperatures up to 1760 K [19]. Such a temperature is not reached even in emergency modes accompanied by failure of emergency protection. Tungsten weakly interacts with lead. The conditions for the formation of lead tungstate ($PbWO_4$) in the reactor core are poor: there is not enough oxygen; high-energy neutrons break chemical bonds [19].

Tungsten is characterized by high thermal conductivity (Fig. 3.8), which increases the efficiency of heat transfer from fuel elements to the coolant. This leads to a decrease in fuel temperature and increased safety in emergency modes with an increase in reactor power.

3.4.6 ATWS Analysis

The design is based on the core layout consisting of fuel assemblies that are geometrically identical to the assemblies of the BREST-OD-300 reactor [36].

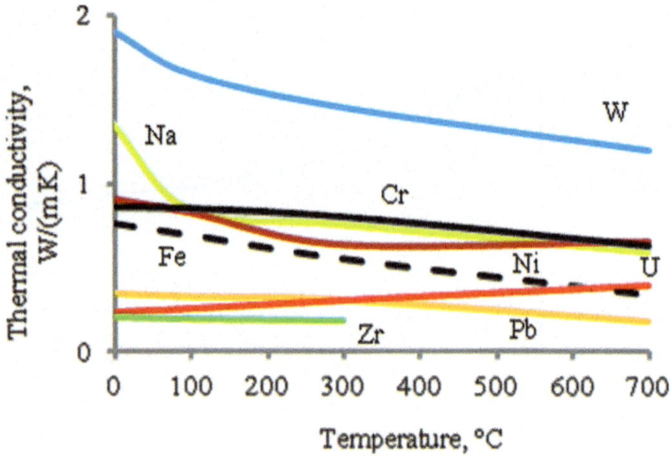

Fig. 3.8 The temperature dependence of the thermal conductivity of some metals (according to [41])

When using conventional materials of the BREST reactor in a reactor with an electric power of 2.4 GW and when using innovative materials (fuel, coolant, fuel pin cladding) allowing to increase the power to 2.9–3.0 GW, the most dangerous is the superposition of three modes from LOF WS, LOHS WS, TOP WS and OVC WS. When four modes are superimposed, any two processes are neutralized. Such superposition is less dangerous than any three of them. The critical element from the point of view of the possibility of ensuring internal self-protection is not exceeding the permissible temperature and strength of the fuel pin cladding. The temperature of the onset of decomposition of nitride fuel (significant increase in gas evolution) is about 2000 K. Calculations show that such temperatures are practically not achieved in ATWS modes. In emergency modes, the melting temperature of metallic uranium nanopowder can be exceeded. But this does not lead to an accident. The determining factor is the melting of nitride in the fuel.

When using innovative cermets fuel, a coolant based on lead and thorium ores (with a ^{208}Pb isotope content in lead of up to 80%), and tungsten fuel cladding, the void reactivity effect is negative (even under the most dangerous scenarios of its development). In all the considered ATWS emergency situations and their combinations, there is a significant temperature margin before lead boils and nitride melts in the fuel. The smallest temperature margin in emergency modes is found in fuel cladding. The maximum cladding temperature is achieved in the LOF WS mode and combinations of this mode with others: (LOF + TOP + LOHS) WS or (LOF + TOP + OVC) WS.

Let us consider the following core variants:

- Initial variant (Variant 1)—MN fuel, natural lead, steel cladding, 2.4 GW /e/, maximum reactivity density effect $\approx 7\ \beta$.

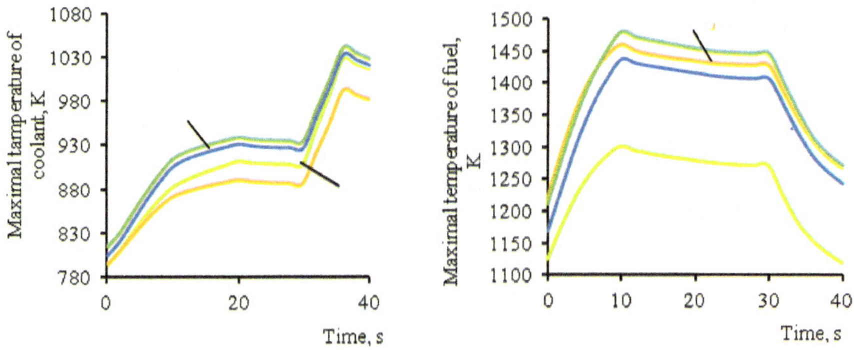

Fig. 3.9 Overlay of TOP WS, LOF WS (at a natural circulation level of 35.24%) and LOHS WS modes for options 1–4: the number next to the curve corresponds to the option number

- Improved variants (2–4). Variant 2—cermets based on UN–PuN micrograins and waste metallic uranium nanopowder, variants 3 and 4—metal ceramics and based on UN–PuN micrograins and enriched (up to 20% vol.) metallic uranium nanopowder, lead-based coolant from thorium ores (with a ^{208}Pb isotope concentration of about 80%) and tungsten cladding.

Figure 3.9 shows the dependences of the maximum temperatures of the fuel and coolant when superimposing the TOP WS, LOF WS (consumption in natural circulation mode is 35.24% of the nominal) and LOHS WS modes for options 1–4. The number near the curve corresponds to the option number.

In this case, it was postulated that all main circulation pumps would be de-energized (coast-down time is equal 3 s), the entire positive reactivity reserve would be introduced (0.99 β in 10 s), and lead would enter the core at a temperature equal to the outlet temperature in the nominal mode (the coolant temperature at the active zone inlet increased in 10 s according to a linear law with time).

Reactor variants 2–4 are reliable and safe in all combinations of emergency situations, including LOCA WS. The maximum void effect of reactivity under the most dangerous scenario is negative. The capacity for variant 2 increases slightly. High-capacity configuration variants: 110% of the nominal (variant 3) and 120% of the nominal (variant 4) are obtained using enriched uranium nanopowder in the fuel. The capacity can be increased by 20% at a ^{235}U concentration in the nanopowder of 14–16%.

If the void effect of reactivity is eliminated by engineering means and reactor design, the flattening of the core can be reduced. When switching to an extended core, the electrical power can be increased to an abnormally high value of 13.3 GW. Reactors of such power will not be in demand in the foreseeable future. For comparison: the largest hydroelectric power station in the world—the Chinese "Three Gorges" has an electric capacity of 22.5 GW. The largest thermal power station in the world is the Chinese "Tuoketuo" with an electric capacity of 6.6 GW, operating on coal. The thermal power station consists of power units with a capacity of 600 MW. The largest thermal power station operating on associated petroleum gas and natural gas is the Russian Surgutskaya GRES-2, consisting of eight power units from 0.40 to 0.83 GW. Its total electric capacity is almost 5.7 GW.

The extended core has an optimal diameter-to-height ratio in terms of minimal neutron leakage. This configuration is a high-temperature reactor. When operating at nominal power, the maximum coolant temperature reaches 1631 K. The use of cermets fuel based on UN–PuN mononitride micrograins and metallic uranium nanopowder, coolant based on lead, thorium ores and tungsten coatings on fuel elements ensures reactor safety. A distinctive feature of the configurations is the reduced hazard of emergency modes when they occur simultaneously. Large reserves of up to the maximum permissible temperatures of the fuel, coolant and structural materials are preserved in all emergency modes. The fuel temperature reaches its maximum value (1878 K) in the TOP WS mode. Superposition of other modes and TOP WS leads to a decrease in the maximum fuel temperature. The maximum coolant temperature for a high-temperature reactor reaches 1897 K when (TOP + LOF) WS is superimposed.

3.5 Discussions

At a temperature of 600–800 °C, superheated water vapor reacts with lead. In this case, free hydrogen is formed and heat is released. The released hydrogen practically does not interact with lead, remaining free. In case of depressurization of the steam generator tubes (the "liquid lead—superheated water vapor" heat exchanger), which is most likely in the LOF WS mode (and superposition of LOF WS with some other LOF WS emergency modes), when the lead temperature is maximum, hydrogen release in the steam generator is possible. Some hydrogen will be drawn into the core together with the steam; some will be separated and will rise upward (against the flow of liquid lead). If a hydrogen explosion occurs in the steam generator with the destruction of the steam generator (or all steam generators), a severe accident will not occur. But if a hydrogen explosion occurs in the upper part of the core, the upper support plate may be torn off together with the devices attached to it, including the automatic reactor shutdown and emergency protection systems. This may lead to a serious accident with the release of radioactive substances outside the nuclear power plant.

The high combustion temperature of hydrogen (3000 °C) ensures the melting of even the most refractory metals, with the exception of tungsten, which melts at a temperature of 3422 °C [41].

Note that in fast reactors with liquid metal coolant, there are no conditions for a hydrogen explosion.

Pure hydrogen does not explode. The danger arises when it comes into contact with atmospheric air. Oxygen is the catalyst for hydrogen explosion. Hydrogen, when mixed with air, forms an explosive mixture—detonating gas. This gas is most explosive when the volume ratio of hydrogen and oxygen is 2:1, or hydrogen and air is approximately 2:5. Mixtures of hydrogen and oxygen of other compositions are also explosive when heated. For hydrogen to explode, an oxygen concentration of 18–59% is required. Despite the need to maintain a certain oxygen concentration in the lead coolant [36], there is clearly not enough oxygen in the circuit of BREST reactors for hydrogen to explode.

The greatest danger is the additional heat release in chemical reactions with hydrogen release. This will lead to further depressurization of an even greater number of steam generator tubes.

If an accident occurs in the core or steam generator with the release of radioactive and (or) stable substances outside the reactor, then hydrogen may be released into the atmosphere and explode.

Thus, to prevent serious accidents in BREST-type reactors with a power significantly greater than BREST-OD-300, for example, BREST-1200 and much greater power, it is necessary to use innovative materials for the core: cermets fuel containing nanopowder of metallic uranium, and a coolant based on lead thorium ores.

Additional experimental justification is required for the use of tungsten coatings of fuel pin cladding. Such coatings may peel off and may be removed from the core. This corresponds to an emergency situation with the introduction of positive reactivity. For this reason and in order to reduce the void effects of reactivity, fuel pin cladding made entirely of tungsten is preferable. To reduce the density and void effects of reactivity required, an additional absorber can be placed in separate tubes in the fuel assemblies of the central part of the core (in the central profiling zone).

The transition to reactors of abnormally high power will require the development of appropriate heat exchange equipment.

3.6 Conclusions

The use of innovative materials in fast energy reactors is due to the predicted mass introduction of high-power reactors into the energy sector by the middle of the twenty-first century.

The physical principles of using innovative cermets fuel with additives of metallic uranium nanopowder are related to the need to increase the average density and thermal conductivity of the fuel to ensure safety in ATWS-type modes. The physical principles of using lead extracted from thorium ores as a coolant are related to the need to minimize the void effect of reactivity in medium- and high-power reactors and limit the risk of TOP WS accidents (increase in the average lifetime of prompt neutrons).

The main task in using innovative materials is to help reduce the coolant temperature in emergency modes and achieve the same Doppler coefficient of reactivity in different emergency situations, including ATWS (for example, in LOF WS and TOP WS). In this case, to improve the self-protection of the reactor, it is necessary to increase the absolute value of the negative Doppler coefficient of reactivity by other means not related to the choice of fuel.

It is preferable to use not one (single) absorber tablet in the upper part of the fuel element of the active zone, but to assemble it from three or four tablets of a smaller height.

It is inappropriate to rely on corrosion and erosion of structural materials as a "burnable" absorber. The reactor campaign increases, but at the same time the probability of damage and destruction of the fuel element cladding increases.

The release of free hydrogen during contact of superheated water vapor with lead coolant or with structural materials is dangerous only if an accident with the release of any substances (and hydrogen) into the atmosphere occurs simultaneously.

There is no need for a free hydrogen neutralizer in steam generators, at the entrance to the core and above the free surface of the lead coolant.

References

1. *The Database on Nuclear Power Reactors* (IAEA Power Reactor Information System). http://www.iaea.org/pris. Accessed 28 Nov 2024
2. *Nuclear Power Reactors in the World, Reference Data Series No. 2* (International Atomic Energy Agency, Vienna, Austria, 2024), 94 pp.
3. *World Survey of Fusion Devices 2022* (International Atomic Energy Agency, Vienna, Austria, 2022), 163 pp.
4. *Scenarios for the Development of Global Energy Up to 2050* (Russian Energy Agency of the Ministry of Energy of Russia, Moscow, 2024), 129 pp.
5. *Energy, Electricity and Nuclear Power Estimates for the Period Up to 2050, Reference Data Series, No. 1* (International Atomic Energy Agency, Vienna, Austria, 2023), 139 pp.
6. *The 17 Goals (Sustainable Development Goals)* (United Nations, Department of Economic and Social Affairs Sustainable Development). https://sdgs.un.org/goals. Accessed 28 Nov 2024
7. V.V. Lemekhov, A.V. Moiseyev, A.A. Bazhanov, M.K. Sarkulov, Current status of the BREST-OD-300 project continuity of solutions for development of commercial lead-cooled reactor facilities. Innovative designs and technologies of nuclear power, in *VI International Scientific and Technical Conference* (2023), p. 28
8. V.N. Polovinkin, *Nanotechnology in Energy* (PRoATOM Agency, 2010). http://www.proatom.ru/modules.php?name=News&file=article&sid=2118. Accessed 28 Nov 2024
9. N.A. Azarenkov, V.N. Voevodin, V.G. Kirichenko, G.P. Kovtun, Nanostructured materials in nuclear power engineering. Bull. Kharkiv Univ. Ser. Phys. Nucl. Part. Fields **887**(1/45), 4–24 (2010)
10. N. Taniguchi, On the basic concept of nano-technology, in *Proceedings of the International Conference on Production Engineering. Part II* (Japan Society of Precision Engineering, Tokyo, 1974), pp. 18–23
11. K.E. Drexler, *Engines of Creation: The Coming Era of Nanotechnology* (Anchor Library of Science, Doubleday, Anchor Books, New York, 1987), 312 pp.
12. K.E. Drexler, *Nanosystems: Molecular Machinery, Manufacturing, and Computation* (Wiley, 1992), 575 pp.
13. ISO, *Technical Committees—TC 229—Nanotechnologies* (London, United Kingdom). https://www.iso.org/committee/381983.html. Accessed 28 Nov 2024
14. L.V. Radushkevich, V.M. Luk'yanovich, On the structure of carbon formed during thermal decomposition of carbon monoxide on an iron contact. J. Phys. Chem. **26**, 88–102 (1952)
15. J.A.E. Gibson, Early nanotubes? Nature **359**(6394), 369 (1992)
16. R. Curl, Jr., Scientists uncover the secret of Damascus steel. NNN—NanoNewsNet, 8 Jan 2008. http://www.nanonewsnet.ru/blog/nikst/uchenye-otkryvayut-tainu-damasskoi-stali. Accessed 28 Nov 2024

17. A.M. Prokhorov, Crucible smelting. Great Sov. Encyclo. **25**(3) (1970–1979). http://bse.uaio.ru/BSE/bse30.htm. Accessed 28 Nov 2024
18. G. Lubin, *Handbook of Composites* (Van Nostrand Reinhold Company Inc., U.S., 1982)
19. V.S. Okunev, Substantiation of the feasibility of using fuel rods with tungsten spraying in power fast reactors of new generation. Therm. Eng. **58**(14), 1167–1171 (2011)
20. O.J. Wick (ed.): *Plutonium Handbook, a Guide to the Technology 2* (American Nuclear Society, 1980)
21. V.S. Okunev, Designing of new generation of the nuclear reactors. AIP Conf. Proc. **2195**, 020012 (2019)
22. S.V. Savushkina, M.N. Polyansky, E.A. Vysotina, A.A. Ashmarin, Formation and study of plasma gradient coating with a top layer of hafnium oxide. Bull. Tomsk Polytech. Univ. Eng. Georesour. **329**(10), 30–39 (2018)
23. S.A. Nikulin, S.O. Rogachev, A.B. Rozhnov, V.M. Khatkevich, T.A. Nechaikina, M.V. Morozov, Structure and properties of a layered steel/vanadium alloy/steel composite prepared by high-pressure torsion. Russ. Metall. (Metally) **2016**(4), 375–379 (2016)
24. M. Tokita, Development of advanced spark plasma sintering (SPS) systems and its applications. Ceram. Trans. **194**, 51–60 (2006)
25. Yu.S. Fedorov, B.E. Burakov, V.M. Garbuzov, A.A. Kitsay, M.A. Petrova, Method for producing nuclear fuel pellets based on uranium dioxide. Patent for invention RU2459289C1, 2012
26. A.F. Grachev, A.A. Maershin, V.N. Golovanov, Experience and prospects of using fuel elements based on vibrocompacted oxide fuel, in *Proceedings of International Science and Technology Conference "Nuclear Power Engineering and Fuel Cycles"*, Moscow, Dimitrovgrad (2003)
27. A.A. Maershin, V.A. Tsykanov, V.N. Golovanov, O.V. Skiba, G.I. Gadzhiev, A.S. Korolkov, N.V. Bobrova, V.A. Kisly, A.A. Teikovtsev, Development and testing of fuel elements for fast reactors with vibration-compacted oxide fuel. At. Energy **91**(5), 378–385 (2001)
28. D.P. Shornikov, M.A. Burlakova, B.A. Tarasov, Obtaining and compacting uranium nitride nanopowders by plasma-spark and electric pulse sintering methods. Vect. Sci. Togliatti State Univ. **3**, 95–98 (2013)
29. *Catalogue of Isotopic Dates of Rocks of the Ukrainian Shield*. Academy of Sciences of the Ukrainian SSR. Institute of Geochemistry and Physics of Minerals. Ministry of Geology of the Ukrainian SSR (Naukova Dumka, Kyiv, 1978), pp. 90–136
30. N.I. Alekseev, S.N. Bol'shagin, E.A. Gomin, S.S. Gorodkov, M.I. Gurevich, M.A. Kalugin, M.S. Yudkevich, The status of MCU-5. Phys. At. Nucl. **75**, 1634–1646 (2012)
31. S. Ganesan, Update of the WIMS-D4 nuclear data library, in *Status Report of the IAEA WIMS Library Update Project, International Nuclear Data Committee, INDC (NDS)-290* (International Atomic Energy Agency, Vienna, 1993), 190 pp. https://www-nds.iaea.org/publications/indc/indc-nds-0290.pdf. Accessed 29 Nov 2024
32. V.S. Okunev, Fundamentally new composite materials of fast reactors made on the basis of nanotechnology. Key Eng. Mater. **887**, 159–164 (2021)
33. V.S. Okunev, New materials and coatings for nuclear technology, in *Proceedings of 7th International Congress on Energy Fluxes and Radiation Effects (EFRE)* (Tomsk, Russia, 2020), pp. 1145–1149. https://doi.org/10.1109/EFRE47760.2020.9242066
34. A.E. Waltar, A.B. Reynolds, *Fast Breeder Reactors* (Pergamon Press, 1981), 853 pp.
35. B.A. Gabaraev, A.I. Filin, Development of a BREST-OD-300 NPP with an on-site fuel cycle for the Beloyarsk NPP. Implementation of the initiative by Russian Federation President V.V. Putin, in *Proceedings of 11th International Conference on Nuclear Engineering, ICONE-11, 2003* (JSME/ASME, Shinjuku, Tokyo, Japan, 2003)
36. V.V. Orlov, *Materials of the Technical Design of the BREST-OD-300 Power Unit* (FSUE "N.A. Dollezhal Research and Development Institute of Power Engineering", Moscow, NIKIET, 2000)
37. K.-H. Neeb, *The Radiochemistry of Nuclear Power Plants with Light Water Reactors* (Walter de Gruyter, Berlin, New York, 1997), 733 pp.
38. L.V. Gurvich, G.V. Karachentsev, V.N. Kondratiev, Yu.A. Lebedev, A. Medvedev, V.K. Potapov, Yu.S. Khodeev, *Chemical Bond Breaking Energies. Ionization Potentials and Electron Affinity* (Moscow, Nauka, 1974), 351 pp.

39. E.A. Orlova, Yu.I. Orlov, E.A. Kryuchkov, V.N. Komyshny, V.S. Egorov, V.G. Zmurin, V.G. Dvortsevoy, S.A. Zagrebaev, K.V. Kolosov, Spontaneous formatted a carbonitride covering on steel from melted eutectic alloy lead-magnesium, in *Third International Scientific and Technical Conference "Innovative Projects and Technologies of Nuclear Power"* (JSC "NIKIET", Moscow, 2012)
40. E.A. Orlova, A.F. Gurbich, S.L. Molodtsov, V.V. Orlov, S.N. Bozin, A.V. Bashkin, Formation and study of protective coatings compatible with nitride fuel on ferritic-martensitic steels. At. Energy **105**(5), 269–274 (2008)
41. E. Lassner, W.-D. Schubert, *Tungsten: Properties, Chemistry, Technology of the Element, Alloys, and Chemical Compounds* (Springer, 1999), p. 443

Chapter 4
Impact Toughness and Tensile Strength Tests of $AlSi_{10}Mg$ Alloy Specimens After Selective Laser Melting

Artyom S. Raznoschikov, Leonid V. Belyaev, Alexey V. Zhdanov, and Nikita S. Dovbysh

Abstract The article presents the results of a study of the mechanical properties of samples obtained by selective laser melting. Impact toughness and tensile strength tests were performed on samples made of $AlSi_{10}Mg$ alloy. Particular attention is paid to comparing the mechanical characteristics of samples printed in different directions. The effect of heat treatment on the strength and impact toughness of samples is also studied. The results of the work can be useful for the development of additive manufacturing and a more comprehensive understanding of the properties of materials.

4.1 Introduction

In the modern world, technologies are developing at an incredible speed, and one of the most promising areas is additive manufacturing. In particular, the method of selective laser melting (SLM) of metal parts from powder materials attracts the attention of researchers and engineers due to its unique capabilities in the field of creating complex products with high precision and unique characteristics [1, 2]. Selective laser melting is an additive manufacturing approach that allows you to create three-dimensional objects by sequentially applying layers of powder material and melting them under the influence of laser radiation. This method allows you to use various initial powder materials, but one of the most popular and in-demand in the industry remains the $AlSi_{10}Mg$ alloy. However, the additive manufacturing method is extremely demanding on the quality and shape of the initial particles of the powder material. Sphericity is often discussed in research, but is rarely determined during quality control of raw materials in real production. Sphericity can be defined as follows: Sphericity = Ideal surface area/Actual surface area [3, 4]. Although it

A. S. Raznoschikov · L. V. Belyaev · A. V. Zhdanov (✉) · N. S. Dovbysh
Vladimir State University named after Alexander and Nikolay Stoletovs, Vladimir, Russia
e-mail: zhav33@mail.ru

© The Author(s), under exclusive license to Springer Nature Switzerland AG 2025
A. Gibadullin et al. (eds.), *Proceedings of the 4th International Scientific and Practical Symposium on Materials Science and Physics*, Springer Proceedings in Physics 318,
https://doi.org/10.1007/978-3-031-94805-3_4

can be calculated, the concept of what "sphericity" is still contains some subjectivity. Roundness and aspect ratio can also be used to describe the sphericity or roundness of a particle. In general, the higher the sphericity, the better the flowability of the powder particles. Better sphericity is expected to create smaller pores during melting. Additive manufacturing requires a sphericity of metal powder greater than 98%, which will facilitate layer deposition and powder flow during production. Another serious defect of parts manufactured by layer-by-layer synthesis on a fine powder substrate is the anisotropy and heterogeneity of the physical and mechanical properties of the material. To test the anisotropy of the properties of the material obtained by SLM, it is necessary to test the hardness of parts and samples in the horizontal and vertical directions at an angle of 45° to the melting plane. Another inherent property of powder spray manufacturing is the presence of trapped gas during the solidification process. It has been noted in [5–7] that gas trapped in the powder particles remains in the printed samples after the additive manufacturing process is complete. The solidification rate is too high for the gas to escape from the particles in 100% of cases. For fatigue-limited parts, this likely means that post-processing will be required, namely a hot isostatic pressing (HIP) cycle to minimize residual pores or defects [8–10]. The SLM method has a number of disadvantages: low productivity, surface quality, porosity, and the need to control mechanical properties. The use of SLM to produce case parts is advisable in several cases: when it is impossible to produce them by other methods (e.g., in the presence of internal conformal channels and voids), a small batch of parts (when the manufacture of a mold for casting is economically unprofitable), and also when the task is to reduce weight by creating a bionic structure. Recently, customers have been making additional demands on manufacturing times while ensuring physical and mechanical properties. As with other methods, the SLM method requires a thorough comprehensive study of the properties of the materials obtained. Papers [11, 12] provide studies of the hardness and tensile strength of the $AlSi_{10}Mg$ material after SLM, but studies devoted to such an important material characteristic as impact toughness are much less common. This characteristic is necessary for parts used in aviation, space and work at negative temperatures. In well-known works on SLM an analysis is made of the effect of subsequent heat treatment modes on the phase composition, porosity, tensile strength, etc. However, there have been no studies on the impact toughness of parts obtained by SLM for this material. Thus, this work is aimed at studying the impact toughness of the $AlSi_{10}Mg$ material after selective laser melting and the effect of subsequent heat treatment on it. At the same time, the paper presents the results of our own tests of tensile strength before and after heat treatment.

4.2 Materials and Methods

4.2.1 Powder Material

At the stage of preparation of the SLM process, it is necessary to have information about the powder material used to build the part: granulometric, chemical and phase composition, thermal conductivity, absorption capacity, melting temperature, solidification, as well as fluidity and bulk density of the powder material. Available data on the powder material allows you to save time on optimizing the parameters of the installation to obtain a defect-free product.

First of all, a comparison was made between the common powder $AlSi_{10}Mg$ with CL 30 Al produced by Concept Laser and the powder of the same alloy RS-300 produced by Rusal. SEM-images of the powder were obtained using equipment Quanta 200 3D (FEI). The powder material RS-300 has a shape close to spherical (Fig. 4.1a) with an average particle size of 30 μm. However, the image shows non-spherical particles and satellites are visible on the surface of the powder granules. The particle shape of the used powder $AlSi_{10}Mg$ (Rusal, RS-300) was compared with the similar composition of the well-known manufacturer of SLM equipment and materials Concept Laser (Germany). SEM-images of powder particles produced in Germany are shown in Fig. 4.1b. The powder composition CL 30 Al also contains non-spherical particles, some particles fused together and satellites.

As noted, in addition to shape defects, powder particles may have internal defects such as pores. In this regard, polished sections of powder material granules were prepared to detect pores that arise during its production. The $AlSi_{10}Mg$ (Rusal, RS-300) powder material was filled with epoxy resin, then ground and polished for further

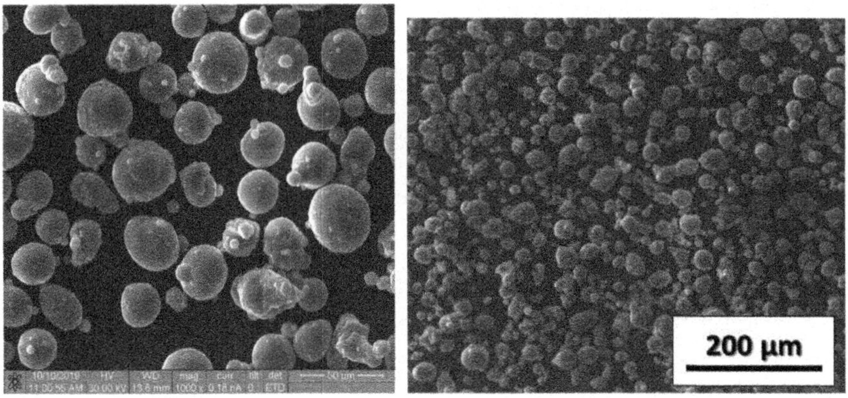

a) RS-300, Rusal, Russia, **b)** CL 30 Al, Concept Laser, Germany

Fig. 4.1 SEM-image of the powder $AlSi_{10}Mg$

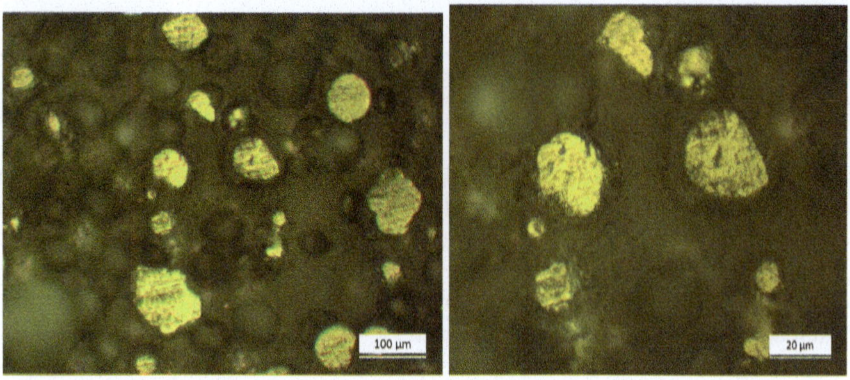

Fig. 4.2 Optical micrographs of $AlSi_{10}Mg$ (Rusal, RS-300) powder sections

study using an optical microscope with Altami MET 1T. Optical micrographs of a section of powder material are shown in Fig. 4.2.

Optical micrographs of sections of powder granules show non-uniform inclusions and, in some places, what can be identified as internal micropores.

In addition, the phase composition $AlSi_{10}Mg$ (Rusal, RS-300) and $AlSi_{10}Mg$ (Concept Laser, CL 30 Al) was measured using a Bruker D8 Advanced X-ray diffractometer, and the elemental composition of the powder was determined using an Olympus Vanta Series C X-ray fluorescence spectrometer.

The results of measuring the phase composition of the initial powders are as follows RS-300/CL 30 Al (in %): Al—94.92/94.8; Fe_2O_3—0.13/0.15; SiO_2—2.32/2.15; Mg_2Si—0.63/0.68; Si—1.99/2.22.

The phase composition of the RS-300 powder contains 0.12% more solid solution of components in aluminum than the CL 30 Al powder. Further, the phase composition of the powders consists of the chemical modifier SiO_2, added as a deoxidizer, its content is 2.05% for the RS-300 powder and 1.94% for the CL 30 Al powder. Also, both powders contain the natural Si O_2 phase, which is 0.27% and 0.21% for the RS-300 and CL 30 Al grades, respectively. Finally, the Mg_2Si phase was found in approximately the same amount in both analyzed powders. The elemental composition of the material used is presented in Table 4.1.

Table 4.1 Results of elemental analysis of powder material $AlSi_{10}Mg$

Mass fraction of elements (%)								
Al	Si	Mg	Fe	Mn	O	Ti	Cu	Zn
$AlSi_{10}Mg$ (Russia) RS-300								
88–91	9–11	0.17–0.3	Up to 1	0.2–0.5	–	–	Up to 0.3	Up to 0.3
$AlSi_{10}Mg$ (Germany) CL 30 Al								
87.2–90.7	9–11	0.2–0.45	0–0.55	0–0.45	0–0.2	0–0.15	0–0.01	0.1

The granulometric composition was studied by scanning electron microscopy using Quanta 200 3D. Most of the powder particles are spherical, the average fraction size is 35 μm, the PC-300 powder has an average fractional composition of 30 μm. The observed grain size distribution falls within the range typically used in materials for SLM manufacturing.

Before the actual sintering process, technological preparation of powder materials is carried out using mechanical mixing and step-by-step sifting, in order to obtain the most uniform granulometric composition.

Despite minor differences in the phase and elemental compositions of the powders, both meet the requirements for use in the SLM method. Further impact toughness and tensile strength tests were performed on samples made of the RS-300 grade powder, since it is more readily available and inexpensive at the moment.

4.2.2 SLM Processing Parameters

The scanning strategy refers to the path the laser beam takes when scanning the powder bed. For each scanning strategy selected, the associated parameters such as laser power and scanning speed should be optimized. In general, high scanning speeds and high laser power lead to high thermal gradients and residual stresses. The optimal combination of laser power, speed and scanning pattern (strategy) determines the success of the SLM process in terms of the achievable result.

When the scanning area is divided into small segments, the laser beam scans successive lines in a very short time, so that the temperature of the already scanned area will still be high when the laser beam scans along another path. In this way, the temperature gradients between adjacent scanning lines are reduced, resulting in reduced thermal stress.

While there are many strategies for building objects in additive manufacturing, during the SLM process, during processing, an island scanning strategy was used, dividing the area into 5 × 5 mm squares (Fig. 4.3). During the processing of each new layer, the scanning direction angle was changed by 45° with an offset in X and Y by 1 mm relative to the previous one.

When interacting with powder metals, the following characteristics of the laser system must be taken into account: laser type, power, wavelength, focal length, beam diameter, etc.

A setup with the following characteristics was used: the wavelength of laser radiation was $\lambda = 1.07$ μm, the layer thickness (t) during production was 30 μm, and the processing was carried out in a nitrogen atmosphere. All the test samples were made at a laser radiation power P = 250 W, a scanning speed of V = 2000 mm/s, and hatch distance h = 105 μm.

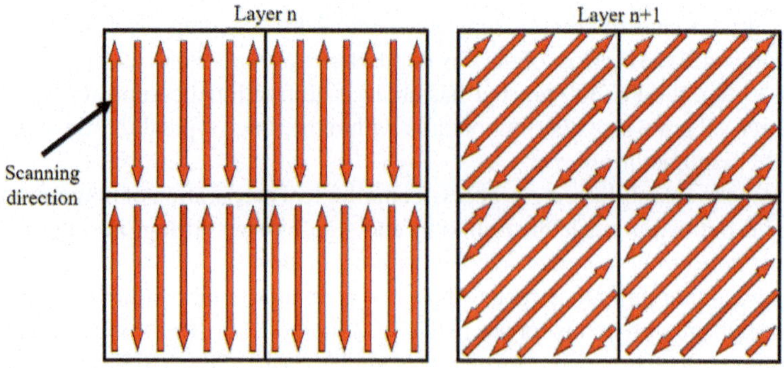

Fig. 4.3 Schematic representation of the scanning strategy used

4.2.3 Sample Form

The shape of the impact toughness samples was regulated by the standard GOST 9454-78 and was 55 × 10 × 10 (L × W × H) mm with a V-type concentrator. The concentrator was implemented after 3D printing, at the sample testing stage. Printed samples for impact toughness are shown in Fig. 4.4a. During SLM, samples were made in horizontal and vertical building direction. For each build direction, 6 samples were made.

The shape of the samples for testing the ultimate strength was regulated by the standard GOST 1497-84. The total length of the specimens was 70 mm, 10 mm at each end with a diameter of 9 mm for fixing in the jaws of the tensile testing machine. The working length was 50 mm with a diameter of 5 mm with fillet radii of 3 mm.

Figure 4.4b shows a photograph of print samples on an aluminum substrate with samples for a tensile testing machine. The photo shows 8 samples obtained in a

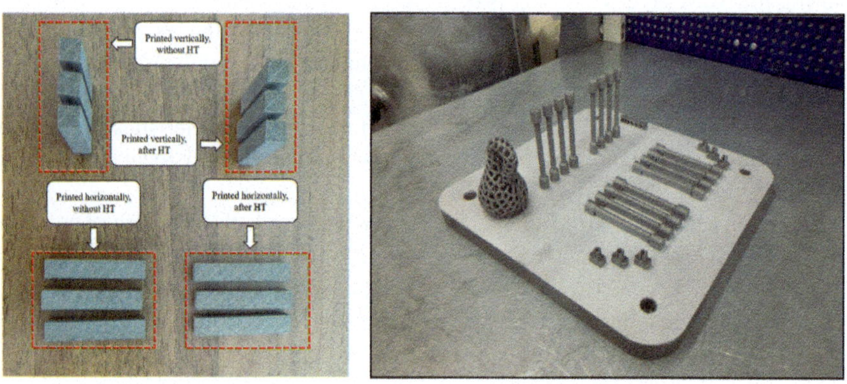

a) for impact toughness
b) for ultimate strength

Fig. 4.4 Photographs of test specimens obtained for vertical and horizontal positions

vertical position and 8 samples in a horizontal position. In the experiments, 6 samples were used, the rest were considered as reserve specimens.

4.2.4 Heat Treatment

Impact toughness and tensile strength tests were performed immediately as build and as build + heat treatment (HT). Heat treatment was performed according to the following regime: heating time 1 h, heating temperature 240 °C, holding time 6 h, subsequent cooling with a furnace to 100 °C.

4.2.5 Microstructure of the $AlSi_{10}Mg$ Alloy After SLM

The samples were separated from the plate using a DK-77-55 electrical discharge machine in the following modes: current of at least 3 A, cutting speed of 20 mm per hour in pulse mode with a current to pause ratio of 16/7. Molybdenum wire with a diameter of 0.18 mm was used. To reveal the microstructure of the material, pre-prepared microsections of the sample were subjected to chemical etching.

Keller's reagent was used for etching, the composition of this reagent includes: HF—1 ml, HCl—1.5 ml, 80 ml of distilled water. The sample was dipped with the polished side into the solution for some time, then washed with running water and dried with filter paper. After the etching procedure, the surface of the sample was studied using a Quanta 200 3D scanning electron microscope.

4.2.6 Mechanical Testing Equipment

Impact toughness tests were conducted using a Metrotest KM-300-M pendulum impact tester at an ambient temperature of 20 °C. Before testing, a B-shaped concentrator of 2 mm depth was formed in the samples, the sample was placed with the concentrator from the pendulum. The tests were carried out using a pendulum with a maximum impact energy of 250 J. Tensile strength was tested on a R-50M tensile testing machine at a temperature of 20 °C.

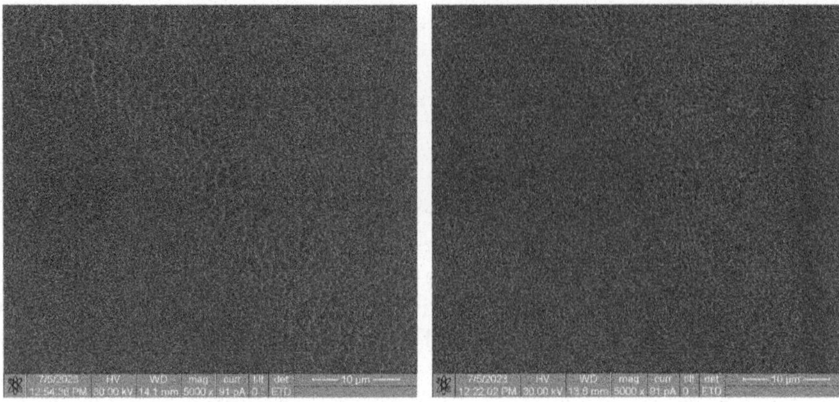

Fig. 4.5 SEM-images of the microstructure of samples

4.3 Results and Discussions

4.3.1 Microstructural Features of the SLM Method

SEM images after polishing and etching of the surface are shown in Fig. 4.5.

The images show a typical SLM microstructure resulting from rapid heating and equally rapid cooling of the material. The images show melt zones and heat-affected zones that arise during SLM processing. The images show areas with a fine-grained structure characteristic of SLM. Also noticeable are areas with a larger structure, apparently formed at the intersections of scanning tracks and reheating, leading to coarsening of the structure.

The heat flow is directed downwards in the SLM process, the highest cooling rate will be at the bottom, which ensures a fine grain structure, with a coarser grain structure observed at the top.

4.3.2 Impact Toughness Test Results

Table 4.2 shows the results of impact toughness tests of samples produced in different growth directions before heat treatment.

The lowest impact strength was found in vertically printed samples. While the horizontal sample show 62.5% higher impact strength with narrower confidence intervals. Table 4.3 are the results of impact strength tests for samples after heat treatment.

Table 4.2 Results of impact strength testing of printed samples as build

No tests	Impact strength without heat treatment (J/cm^2)	
	Vertically	Horizontally
1	3.89	4.75
2	2.86	5.22
3	2.86	5.69
Average value	3.2	5.2
Confidence interval	3.2 ± 0.5545	5.2 ± 0.43

Table 4.3 Results of impact strength testing of printed samples after HT

No tests	Impact strength after HT (J/cm^2)	
	Vertically	Horizontally
1	3.37	4.35
2	2.64	5.05
3	3.49	5.11
Average value	3.2	4.8
Confidence interval	3.2 ± 0.43	4.8 ± 0.3847

Thus, the applied heat treatment did not change the average impact strength for the vertically printed samples, but decreased the impact strength of the horizontal samples on 7.7%. At the same time, the heat treatment narrowed the confidence interval for horizontally and vertically manufactured samples, and the vertical samples still showed significantly lower impact strength. In general, the decrease in impact strength on horizontal samples can be explained by the different thermal exposure times when printing vertical and horizontal samples. Vertical samples consist of more layers than horizontal ones, and the processing of the next thermal layer also affects the previous already fused layers.

4.3.3 Tensile Strength Tests

Tables 4.4 and 4.5 present data on the tensile strength of the tested samples. During the work, 3 samples printed in two different directions were tested, and the average tensile strength and confidence interval were calculated.

From the obtained results it can be observed that the samples in the horizontal building direction without subsequent heat treatment show a tensile strength greater by approximately 15–20 MPa. Generally, stress relieving heat treatment improves ductility but also reduces the ultimate strength of the material. The detailed data are given in Table 4.5.

Table 4.4 Results of tensile strength testing of printed samples without heat treatment

No tests	Tensile strength as build (MPa)	
	Vertically	Horizontally
1	398.72	416.24
2	403.20	415.27
3	405.90	423.42
Average value	402.60	418.31
Confidence interval	402.6 ± 3.34	418.31 ± 4.10

Table 4.5 Results of tensile strength testing of printed samples after HT

No tests	Tensile strength after HT (MPa)	
	Vertically	Horizontally
1	376.01	393.33
2	376.98	383.40
3	375.25	385.84
Average value	375.25	387.52
Confidence interval	375.25 ± 0.80	387.52 ± 4.78

Analysis of Tables 4.4 and 4.5 shows that in the heat-treated samples, there is a decrease in Tensile strength by 6.8% for vertical samples and by 7.4% for horizontal samples.

The phase composition of the obtained SLM samples was studied. During the powder melting process, the phases underwent changes: Al—93.67%; Si—1.19%; Mg_2Si—0.71%; SiO_2—2.99%. In quantitative terms, a decrease in the Al phases by 0.14% and Si by 0.8% is observed with an increase in the mass fraction of the Mg_2Si phases by 0.08% and SiO_2 by 0.67%. Obviously, this fact indicates some decrease in the mechanical properties of the annealed samples.

Figure 4.6 shows the load-deformation graphs of samples grown at different orientations relative to the build platform during tensile testing of the samples.

The tensile strength after heat treatment decreases for both growth directions by an average of 7%. At the same time, the tensile strength in the horizontal direction still remains higher by approximately 10 MPa. At the same time, the subsequent heat treatment probably affected the confidence interval, since the spread of the tensile strength values of the vertically printed sample became significantly smaller. However, this did not happen with the horizontally printed sample.

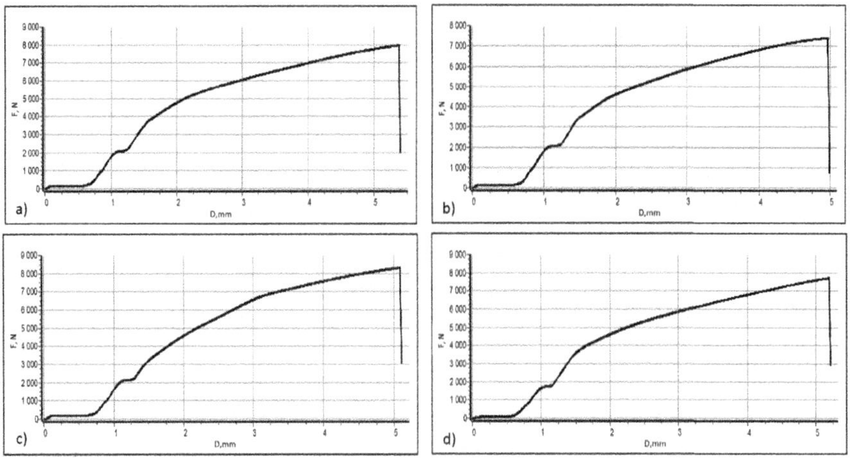

Fig. 4.6 Load-deformation diagrams of samples during tensile testing: **a** vertically printed sample without heat treatment; **b** vertically printed sample after heat treatment; **c** horizontally printed sample without heat treatment; **d** horizontally printed sample after heat treatment

4.4 Conclusion

During the work, a comparison of the particle shape, phase and elemental compositions of the powders of the $AlSi_{10}Mg$ alloy from two manufacturers was carried out. Despite the fact that the powders are almost identical, with the exception of small differences in phase and elemental compositions, further 3D printing of samples was carried out from the more accessible RS-300 powder. Using a scanning electron microscope, data on the microstructure of samples synthesized from the powder material used were obtained. Samples were made for testing impact toughness and tensile strength in the vertical and horizontal directions. A comparison of these characteristics was carried out depending on both the printing direction and the presence of subsequent heat treatment. As a result of the work, the following results were obtained:

- In impact toughness tests, horizontally printed samples show 62.5% higher values than vertical samples;
- Heat treatment did not significantly change the average impact strength for samples printed in both directions;
- When testing the tensile strength of horizontally building samples, this indicator is 10–20 MPa higher than that of vertical ones, both before and after heat treatment;
- The heat treatment used reduces the tensile strength for both growing directions by an average of 7%.

In the future, tests of the strength limit and impact toughness at temperatures different from the environment are planned. First of all, at negative temperatures, since the alloy used is often used in the aviation industry and the parts used are often operated at negative temperatures.

Acknowledgements This work was supported by RSF 23-29-00545.

References

1. M. Ahmadi, Review of selective laser melting of magnesium alloys: advantages, microstructure and mechanical characterizations, defects, challenges, and applications. J. Mater. Res. Technol. **19**, 1537–1562 (2022)
2. J. Gunasekaran, P. Sevvel, J. Solomon, Metallic materials fabrication by selective laser melting: a review. Mater. Today **37**, 252–256 (2021)
3. L. Haferkamp, L. Haudenschild, A. Spierings, K. Wegener, K. Riener, S. Ziegelmeier, G.J. Leichtfried, The influence of particle shape, powder flowability, and powder layer density on part density in laser powder bed fusion. Metals **11**(3), 418–422 (2021)
4. B. Xie, Y. Fan, S. Zhao, Characterization of Ti6Al4V powders produced by different methods for selective laser melting. Mater. Res. Express **8**(7), 076–081 (2021)
5. R. Hebert, Y. Sun, M. Aindow, E. Garboczi, Three-dimensional particle size, shape, and internal porosity characterization: application to five similar titanium alloy (Ti–6Al–4V) powders and comparison to two-dimensional measurements. Addit. Manuf. **44**, 122–129 (2021)
6. S. Lowell, J. Shields, *Powder Surface Area and Porosity* (Springer Science & Business Media, 2013)
7. S. Shrestha, T. Starr, K. Chou, Porosity analysis in metal additive manufacturing by micro-CT, in *ASME International Mechanical Engineering Congress and Exposition 52019* (2018), pp. 1025–1028
8. A. Voznesenskaya, A. Zhdanov, S. Raznoschikov, Evolution of porosity depending on SLM mode and subsequent HIP processing. J. Phys. Conf. Ser. **2077**(1), 012023 (2021)
9. T. Hirata, T. Kimura, T. Nakamoto, Effects of hot isostatic pressing and internal porosity on the performance of selective laser melted $AlSi_{10}Mg$ alloys. Mater. Sci. Eng. A **772**, 138–145 (2020)
10. L. Sun, Y. Yang, S. Li, W. Chen, Y. Wang, P. Yan, B. Hu, Influencing mechanisms of hot isostatic pressing on surface properties of additively manufactured $AlSi_{10}Mg$ alloy. J. Mater. Process. Technol. **329**, 118–125 (2024)
11. J. Chen, Microstructure, porosity and mechanical properties of selective laser melted $AlSi_{10}Mg$. Chin. J. Aeronaut. **33**, 2043–20547 (2020)
12. A. Babaytsev, A. Orekhov, L. Rabinskiy, Properties and microstructure of $AlSi_{10}Mg$ samples obtained by selective laser melting. Nanosci. Technol. Int. J. **11**(3), 11–17 (2020)

Chapter 5
Physicochemical Preparation for Fabrication and Characterization of Gold Nanoparticles for Antibacterial Activity Study

Saif Al Busaidi, Geetha Devi, and Qais Mohammed Mahmood Al Balushi

Abstract It is substantiated that nanoparticles have attracted much attention in many fields due to their extremely small size, geometry, monodispersity, non-toxicity and attractive structural and morphological features. Gold nanoparticles (Au-NPs) are a class of versatile nanomaterials used in all fields. In this study, the physicochemical analysis of Au-NPs, which are prepared from *Azadirachta indica* (neem leaf) extracts and aqueous chloroauric acid by biosynthesis followed by stabilization and centrifugal separation at ambient temperature, is carried out. The fabricated Au-NPs were used in the study of the antibacterial activity of Gram-negative bacteria using radial disk diffusion assay. The progress in the development of nanoparticles is monitored using ultraviolet spectroscopy, Fourier Transform Infrared, X-ray Diffraction, Energy Dispersive X-ray analysis and scanning electron microscopy. The characterization results confirm the successful synthesis of nanoparticles with high yield and well-structured surface morphology and elemental composition, which is confirmed by Scanning Electron Microscopy Image and Energy Dispersive X-ray spectra.

5.1 Introduction

Nanotechnology and novel material development is a very interesting theme of scientific research with enormous scope in electronics, biomaterials, medicine, and energy storage production, etc. Development of harmless and environmentally safe nanoparticles is a paramount priority in the emerging research. Gold nanoparticles are widely employed in drug delivery, medical imaging, preliminary diagnosis, oncology, fabrication and development of innovative biomedical products owing to their outstanding functional properties. Nano materials are chosen in a variety of applications due to

S. Al Busaidi · G. Devi (✉) · Q. M. M. Al Balushi
College of Engineering, National University of Science and Technology, Muscat, Sultanate of Oman
e-mail: geethadevi@nu.edu.om

© The Author(s), under exclusive license to Springer Nature Switzerland AG 2025
A. Gibadullin et al. (eds.), *Proceedings of the 4th International Scientific and Practical Symposium on Materials Science and Physics*, Springer Proceedings in Physics 318,
https://doi.org/10.1007/978-3-031-94805-3_5

their attractive surface and photosensitive properties [1]. Au-NPs have been extensively applied in pharmaceutical and biomedical applications due to their appealing surface structural properties [2]. The excellent surface active properties contribute the drug binding capability of Au-NPs, antibodies, protein, and oligonucleotides [3]. The optical properties of Au-NPs play a prominent role as bio imaging and bio marking agent. The green techniques have been suggested as an auxiliary method for the production of various nanoparticles. Plant based and biological synthesis techniques utilize ecofriendly materials such as bacteria, fungus, and plant extract. Natural plant extracts have been investigated to accomplish the biogenic reduction of the metal, thus act as capping and stabilizing agents [4]. Plant based extracts normally contains carbohydrates, green terpenoids, polyphenols, alkaloids, phenolic acids, and proteins which are exceptional reducing agents; these are much suitable for the production of silver and gold nanoparticles [5]. Numerous synthesis routes have been explored in the preparation of metal nanoparticles using cinnamon, *Hibiscus cannabinus* leaf, henna, and curcumin [6–8]. These materials have simple synthesis methods and also exhibit boundless chemical as well as thermal stability [9]. One such method is the fabrication of AuNPs using biological systems such as fungi, microbes, and plant extracts [10–12]. Preparation of AuNPs via green synthesis route is gaining considerable attention due to its easy and fast processing, cost-effectiveness, environmentally friendly and sustainable nature [13].

The unique features of AuNPs have fascinated particular attention in biomedical field due to their no toxic nature. Majority of the techniques employed in the synthesis of nanoparticles are chemical, physical and biological methods and these methods are much expensive and lead to harmful environment. By considering these limitations, this research recommends a green and inexpensive method to synthesize AuNPs from natural source such as neem leaf extracts (*Azadirachta indica*) by biosynthesis methods. Due to the extremely small size, high surface to volume ratio, prolonged stability, non-toxicity and tunable physical and chemical properties of AuNPs are emerging as a promising drug carrier for cancer treatment.

Very few studies have been carried out on the development AuNPs from plant based materials. The main objective of this research was to develop a cost-effective and environmental friendly approach to synthesize AuNPs in a more sustainable way. This research project aims to integrate nanotechnology and medicine utilizing natural resource to synthesize AuNPs from plant *A. indica* extracts.

5.2 Materials and Methods

The materials used for the synthesis of nanoparticles are Neem leaf extracts (*A. indica*), Chloroauric acid ($HAuCl_4$) and Distilled Water. The methodology employed in the preparation of AuNPs is biosynthesis process. The antimicrobial activity of the synthesized nanoparticles is carried out by radial disc diffusion assay. Figure 5.1 illustrates an overview of various steps involved in the synthesis of AuNPs from neem leaves.

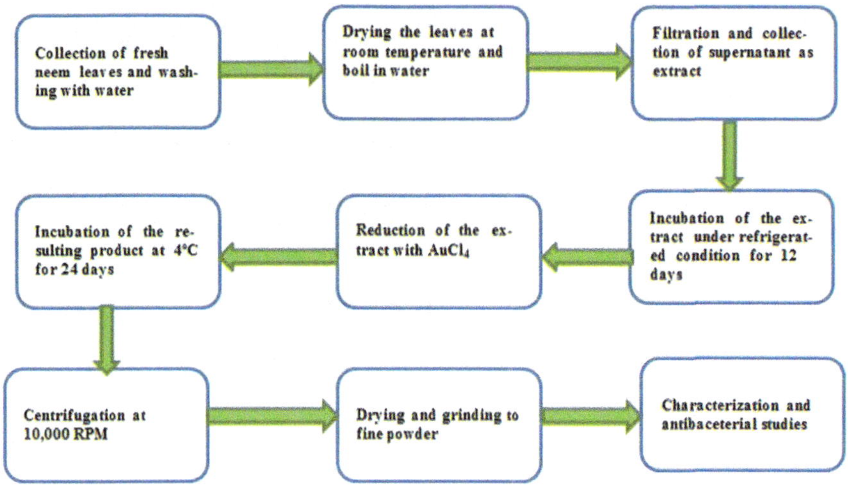

Fig. 5.1 Sequence of steps involved in the synthesis of AuNPs

Stage 1 Synthesis of AuNPs

The synthesis of AuNPs was performed by incubating the plant extract with aqueous gold ions in which the plant extract acts as a reducing agent. The neem extract was prepared by weighing 100.0 g of the fresh neem leaves and boiled in 1000 ml distilled water for a duration of 50 min. The reduction reaction was carried out by mixing 5.0 ml of the plant extract with 45.0 ml of 1 mM $HAuCl_4$ solution. The $AuCl_4$ reduction was constantly examined and the absorbance values are recorded at definite time duration. After reaction, the resulting mixture was centrifuged to 10,000 rpm for 10 min to separate AuNPs followed by multiple washing steps and then dried in an oven at 110 °C followed by characterization studies. The resulting product was kept in a refrigerator for 24 h. Figure 5.2 represents the image of neem leaves and the neem leaf extract after 24 h of incubation is shown in Fig. 5.3. Figure 5.4a–d represents the various stages of nanoparticle synthesis steps. Finally the dried nanoparticle samples were milled into fine powder followed by material characterization.

Stage 2 Characterization

The characterizations tools used for the analysis of AuNPs are UV–visible spectroscopy, Scanning Electron Microscopy, X-Ray Diffraction, Energy Dispersive X-Ray, and Fourier Transform Infrared.

Stage 3 Antimicrobial activity studies

The synthesized AuNPs were employed in the antimicrobial activity using a Gram-negative bacterium.

Fig. 5.2 Dried neem leaves

Fig. 5.3 Neem leaf extract after 24 h of incubation

5.3 Results and Discussions

The morphology of the green synthesized AuNPs are performed using Scanning Electron Microscopy. The microstructural analysis of the nanoparticles are captured at a magnification of 100× and an excitation voltage of 20 kV as shown in Fig. 5.5. As seen in Fig. 5.5, the AuNPs are well formed and evenly distributed with less aggregation confirms the successful preparation of nanoparticles. The absorbance spectra of the fresh extract and the extract after 12 days are indicated in Fig. 5.6. The formation of gold nanoparticles was confirmed by observing the colour change. The resulting AuNPs suspension was separated by centrifugation at 10,000 rpm for 20 min and repeated washing. Supernatant was discarded and the particles settle at the bottom of the tube was washed several times with deionised water.

The progress of particle formation was confirmed using UV–Spectroscopic analysis of the solution at different absorbance values as shown in Fig. 5.6. The change in colour from light yellow to ruby red was observed during the reduction reaction, which indicates the formation of gold nanoparticles. The appearance of peaks mainly depends on the incubation time and wavelength. These results are in agreement with the reported data from previous studies [14, 15].

Fig. 5.4 Experiment results: **a** reaction mixture after 35 min; **b** reaction mixture after 45 min; **c** reaction mixture after 12 h of incubation; **d** final product after 24 h

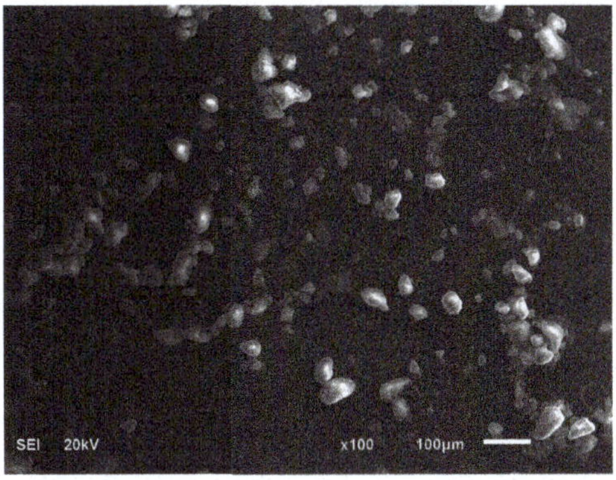

Fig. 5.5 Scanning electron microscopy image of AuNPs

The functional groups present in the AuNPs sample was analyzed using FT-IR spectroscopy with a wave numbers ranges between 4000 and 400 cm^{-1}. Figure 5.7 shows the FT-IR spectra of nanoparticles. The C=C band was observed at a wavenumber of 1624 and 3460 cm^{-1} in the spectrum indicates the presence of O–H bond. The detection of the C=C bond and the O–H bond in the AuNPs shows the

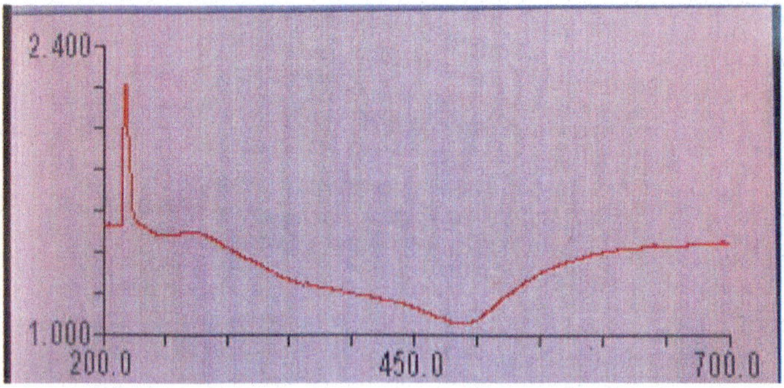

Fig. 5.6 UV–visible absorbance spectra of the extract after 12 days

presence of terpenoids that are present in the neem leaf extract. The FT-IR spectrum of AuNPs showed a very strong peak at 3463 cm^{-1} was assigned as –NH stretching of primary amine band. The stretching vibration displayed at wavenumber between 3000 and 3500 cm^{-1} designates the existence of O–H stretching, which signifies the existence of flavonoids, phenols, benzophenones, flavonoids, and anthocyanins [16, 17]. The infrared spectral distribution of AuNPs are displayed in Fig. 5.8.

The X-Ray diffraction technique was used to identify the crystal structure and phases of the nanoparticles. It was found that the Au-NPs were crystalline in nature with cubic lattice. Powder X-ray diffraction pattern in Fig. 5.9 indicates the crystalline nature of Au-NPs.

The antibacterial assay of Au-NPs: The Au-NPs were tested for the antibacterial activity by disc diffusion method [18]. The stock solution was diluted to different concentrations ranging from 2.5 to 100 μl/ml. From the diluted sample, 10 μl was permeated into a sterile blank disc and 5 μl of Au-NPs suspension was spotted alternately on both sides of the discs and allowed to dry and inoculated with agar media. Table 5.1 indicates the results of antibacterial activity study of the synthesized gold

Fig. 5.7 Fourier transform infrared spectrum of extract

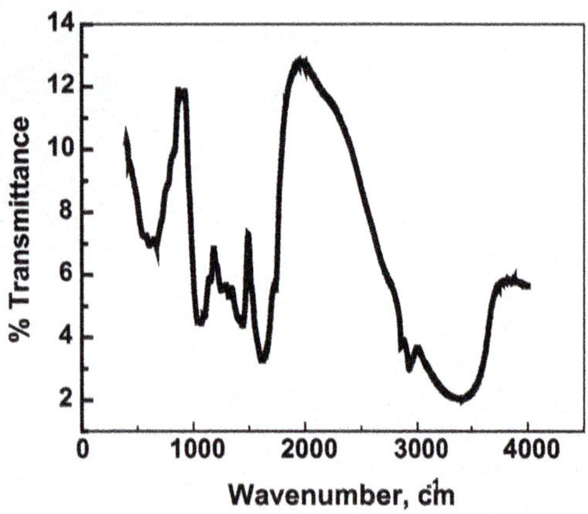

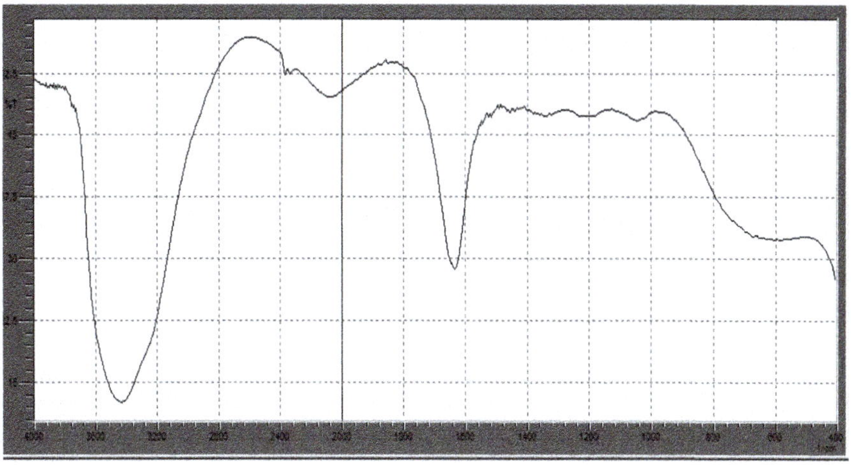

Fig. 5.8 Fourier transform infrared spectrum of AuNPs

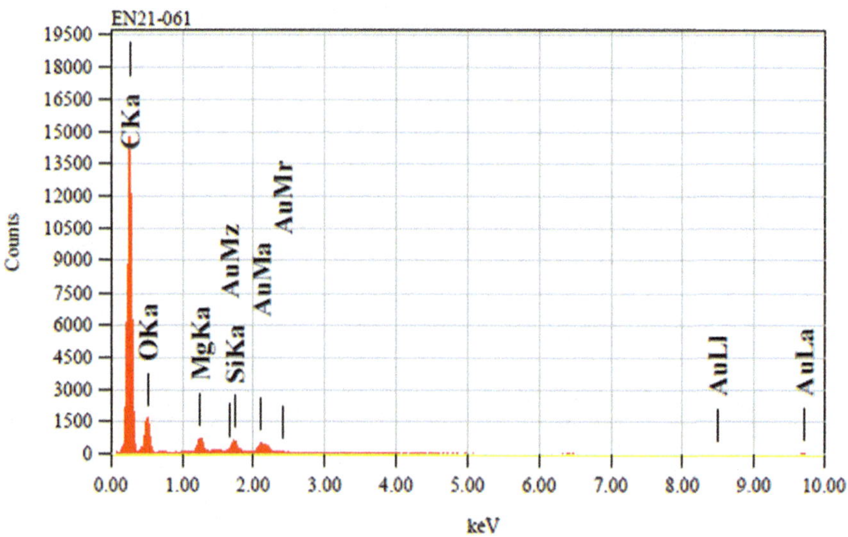

Fig. 5.9 X-ray diffraction spectra of Au-NPs

nanoparticles. The antibacterial valued displayed in Table 5.1 shows the effectiveness of the nanoparticles in destroying pathogens.

Table 5.1 Antibacterial activity study of the Au-NPs

Sample	Sample name	Value (cfu/ml)
1	Fresh neem extract	287×10^5
2	Neem extract after 12 days	409×10^4

5.4 Conclusion

In this study, Au-NPs are successfully biosynthesized from neem leaves by ecofriendly technique and to assess its antibacterial activity using gram positive bacteria. The outcome of the study demonstrates that *A. indica* is an excellent capping and reducing agent in the synthesis of Au-NPs. The UV-Spectroscopy, Fourier Transform Infrared, Scanning Electron Microscopy, and X-Ray Diffraction analyses confirmed the successful synthesis of Au-NPs. Au-NPs are considered as one of the most interesting types of nanostructured material applied in antimicrobial, electronic, catalytic, and various biomedical applications. The values obtained from the study shows the excellent performance of neem leaf extract in antibacterial activity. This cost-effective and environmental friendly method will open up new avenues in synthesizing Au-NPs of size less than 50 nm in a more sustainable and economical way. This simple, single step eco-friendly procedure in the synthesis of Au-NPs appears to be suitable for large scale production as it is economical, rapid, and environmentally benign and is safe for clinical research.

Acknowledgements The research team would like to express their sincere thanks to **National University College of Engineering** for awarding the research project and providing the infrastructure facilities to undertake this project under the **National University Student Research Grant** (**Ref. No. NU SRG 19.3**).

References

1. F. Abdulwahab, F.Z. Henari, S. Cassidy, K. Winser, Synthesis of Au, Ag, curcumin Au/Ag, and Au-Ag nanoparticles and their nonlinear refractive index properties. J. Nanomater. **3**(2), 23–25 (2016)
2. O.S. El Mitwalli, O.A. Barakat, R.M. Daoud et al., Green synthesis of gold nanoparticles using cinnamon bark extract, characterization, and fluorescence activity in Au/eosin Y assemblies. J. Nanopart. Res. **22**, 309 (2020)
3. A. Chahardoli, N. Karimi, F. Sadeghi, A. Fattahi, Green approach for synthesis of gold nanoparticles from *Nigella arvensis* leaf extract and evaluation of their antibacterial, antioxidant cytotoxicity and catalytic activities. Artif. Cells Nanomed. Biotechnol. Int. J. **46**(3), 579–588 (2018)
4. X. Geng, T.Z. Grove, Repeat protein mediated synthesis of gold nanoparticles: effect of protein shape on the morphological and optical properties. RSC Adv. **5**(3), 2062–2069 (2015)
5. V.V. Makarov, A.J. Love, O.V. Sinitsyna, S.S. Makarova, I.V. Yaminsky, M.E. Taliansky, N.O. Kalinina, Green nanotechnologies: synthesis of metal nanoparticles using plants. Acta Nat. **6**(1), 35–44 (2014)

6. S. Menon, S. Rajeshkumar, S. Venkat Kumar, A review on biogenic synthesis of gold nanoparticles, characterization, and its applications. Resour.-Effic. Technol. **3**, 516–527 (2017)
7. H. Nadaroğlu, A. Alayli Güngör, S. İnce, Synthesis of nanoparticles by green synthesis method. Res. Rev. **1**(1), 6–9 (2017)
8. N. Naseri, E. Ajorlou, F. Asghari, Y. Soltanahmadi: Artif. Cells Nanomed. Biotechnol. **46**, 1111–1121 (2018)
9. K.J. Rao, S. Paria, *Aegle marmelos* leaf extract and plant surfactants mediated green synthesis of Au and Ag nanoparticles by optimizing process parameters using Taguchi method. ACS Sustain. Chem. Eng. **3**(3), 483–491 (2015)
10. F. Shahidi, A. Hossain, Bioactives in spices, and spice oleoresins: phytochemicals and their beneficial effects in food preservation and health promotion. J. Food Bioact. **3**, 8–75 (2018)
11. K.H. Smila, T.P. Priya, N.C. Saraswathy, D. Yuvaraj, M. Chandran, P. Vivek, Green synthesis of gold nano particles from cassia auriculata leaf aqueous extract and its cytotoxicity effect on in vitro cell line. Int. J. ChemTech Res. **6**(9), 4241–4250 (2014)
12. D. Son, Z. Bao, Nanomaterials in skin-inspired electronics: toward soft and robust skin-like electronic nanosystems. ACS Nano **12**, 11731–11739 (2018)
13. Y. Xia, H. Yang, C.T. Campbell, Green chemistry principles and practice. Acc. Chem. Res. **46**(8), 1671–1672 (2013)
14. M. Yazdankhah, H. Veisi, S. Hemmati, In situ immobilized palladium nanoparticles (Pd NPs) on fritillaria imperialis flower extract-modified graphene and their catalytic activity for reduction of 4-nitrophenol. Taiwan Inst. Chem. Eng. **91**, 38–46 (2018)
15. K.T. Yong, M.T. Swihart, H. Ding, P.N. Prasad, Preparation of gold nanoparticles and their applications in anisotropic nanoparticle synthesis and bioimaging. Plasmonics **4**(2), 79–93 (2009)
16. Q. Zhang, X. Yang, J. Guan, ACS Appl. Nano Mater. **2**, 4681–4697 (2019)
17. Y.A. Balakin, K.B. Yunusov, A.N. Khaulin, Improvement of heat-resistant steel manufacturing adaptability by combined treatment. Chem. Pet. Eng. **52**(9–10), 717–720 (2017)
18. M. Ahmed, D.W. Pan, M.E. Davis, Lack of in vivo antibody dependent cellular cytotoxicity with antibody containing gold nanoparticles. Bioconjug. Chem. **26**(5), 812–816 (2015)

Chapter 6
Application of the Metaheuristic Algorithms of Global Optimization for Determination of the Composite Materials Constituent's Properties

Andrei Panteleev, Nikolay Turbin, and Maksim Mukhai

Abstract A mathematical model has been formed for calculating the elastic properties of a transversely isotropic fiber and an isotropic matrix based on the characteristics of a composite monolayer. The problem of choosing the parameters of materials is formulated as the problem of minimization of the objective function on a parallelepiped set of admissible solutions. Two methods of global optimization are applied for solution: sine–cosine method and gaining-sharing knowledge-based algorithm, both of which belong to metaheuristic group of optimization methods. The applied problem solution is obtained. The constituent's properties of composite materials were found.

6.1 Introduction

Design and fabrication of composite structures is impossible without information about elastic properties of the materials. At the same time, the constituent properties of the composite material may differ significantly from composite material itself. Due to difficulties in evaluation of constituent properties experimentally, analytical and numerical methods were developed to accomplish this task. One of the most used methods for estimating the properties of composites is micromechanical modeling, where the structural peculiarities of the composite observed at the micro scale are taken into account for prediction of the effective elastic properties of composite material at monopoly scale [1].

Figure 6.1 demonstrates schematically the representative volume element (RVE) of the phase change composite material (PCM) at micro scale, which consists of several components (in the simple case, the constituents are only matrix and fiber). Fiber is a reinforcement of the material, because of its dominant strength and stiffness

A. Panteleev (✉) · N. Turbin · M. Mukhai
Moscow Aviation Institute (National Research University), Moscow, Russia
e-mail: avpanteleev@inbox.ru

© The Author(s), under exclusive license to Springer Nature Switzerland AG 2025

A. Gibadullin et al. (eds.), *Proceedings of the 4th International Scientific and Practical Symposium on Materials Science and Physics*, Springer Proceedings in Physics 318,
https://doi.org/10.1007/978-3-031-94805-3_6

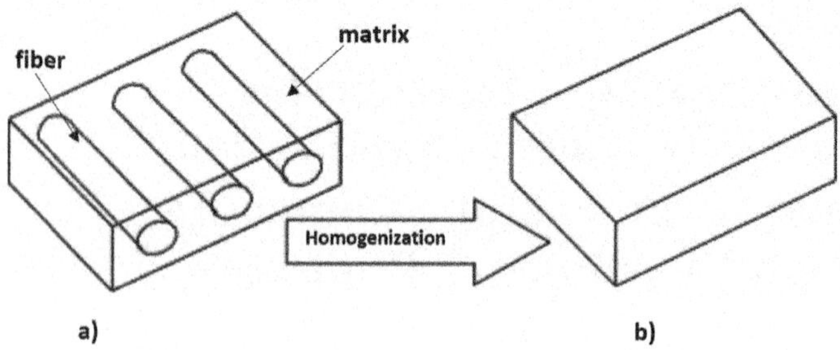

Fig. 6.1 Homogenization process. **a** Heterogeneous material, **b** equivalent homogeneous material. Processed from [2]

with respect to matrix. Matrix, in turn, act as a connecting medium which integrates the fibers in a single substance. Together, matrix and fiber comprise a continuum material with anisotropic properties (anisotropic material has variable properties in different directions in contrast to isotropic material, which has the same properties in all orientations).

Homogenization [2] is a mathematical approach for determination of the effective properties of materials with heterogeneous structure. In a particular case, this method is used for the analysis of elastic properties of the materials, which consist of components with different respective elastic properties. Homogenization might imply the periodicity of internal structure of the material at micro scale, which alleviates the prediction of accurate mechanical properties of the equivalent material at macro scale.

Homogenization, based on analytical techniques, is a convenient tool, capable for integration in optimization process of the design and manufacturing of new composite materials. The analytical approaches for elastic properties homogenization of heterogeneous materials are well developed; basic of them could be found elsewhere [3–9]. Such methods aim for improving the accuracy of the properties prediction. Results of the analytical homogenization were shown to have upper and lower bounds of values, which might be determined by variational principles [4] and depend mainly on the volume fractions of the components and their physical properties. Most of the methods are based on the solution for equivalent eigenvalues deformation, developed in [8, 9] for the case of the elliptic inclusion, embedded in the infinite medium. Self-consistent method [6, 7] is another example of widely implemented methods of homogenization, in which random distribution of elastic inclusions in infinite medium is considered. Methods based on solutions for periodic structures have also achieved great interest [10–12].

The direct problem of homogenization of the monopoly of carbon fiber reinforced plastic (CFRP) consists in solution for equivalent properties of generalized material, when properties of its constituents (transversely isotropic fiber and isotropic matrix) are known. This way, when fiber and matrix properties are given, one can predict the properties of the monolayer containing different combinations of components, e.g. different volume fractions of constituent materials comprising monolayer.

Meanwhile, in practice necessary properties of matrix and fiber are usually unknown or hard to be evaluated experimentally. That is the reason, why inverse problem of homogenization is relevant: determine the properties of fiber and matrix using analytics after properties of the monolayer known from experiment.

For the solution of the inverse problem of homogenization, the formulation of the parametric optimization task is suggested as follows. First, the characteristic parameters of the matrix and fiber, that define the results of the homogenization, are bounded in the intervals of admissible values using a priori information. The objective function is set in the form of maximum cumulative deviation of the predicted monolayer properties from experimentally measured. The solution of the constrained optimization problem is conducted using metaheuristic algorithms, which results in the approximate solution with reasonable accuracy and efficiency in terms of solution time [13–15]. Metaheuristic algorithms are efficiently implemented in the problems of parametric optimization of engineering systems [16–25]. For the optimization task at hand two methods are proposed: Sine–Cosine Algorithm (SCA) [26, 27], which employs the properties of the harmonic functions in the solution search process strategy, and the method of imitating the process of gaining and sharing knowledge among people, known as Gaining-Sharing Algorithm (GSA) [28].

The present investigation is organized in the following way. In Sect. 6.2.1 the formation process for homogenization mathematical model is described. Section 6.2.2 contains formalization of the solution process for prediction of the properties of matrix and fiber in the constrained optimization problem form. The step-by-step algorithm of objective function evaluation is proposed. The strategies of the two implemented methods of optimization are discussed. The algorithm for calculating the objective function value is described in Sect. 6.2.3. Section 6.3 provides the solution of the applied inverse problem of homogenization for composite material using it's experimentally obtained properties. Section 6.4 contains the conclusions made.

6.2 Materials and Methods

6.2.1 The Formation Process for Homogenization Mathematical Model

Input parameters for homogenization procedure are elastic properties of the monolayer material components.

Elastic properties. In Hooke's law, the elastic modulus equals the quotient of applied stress σ and strain ε: $E = \frac{\sigma}{\varepsilon}$. For example, using notation of the material coordinate system on Fig. 6.2, $E_1 = \sigma_1/\varepsilon_1$, where 1 is a direction along fibers. Poisson's coefficient is defined as the quotient of transverse strain and the strain along the loading direction (Fig. 6.2): $\nu_{12} = \frac{\varepsilon_2}{\varepsilon_1}$, when the material is loaded along fibers, i.e. in direction 1. For CFRP, elastic properties of the components are the elastic properties of transversely-isotropic fiber and isotropic matrix.

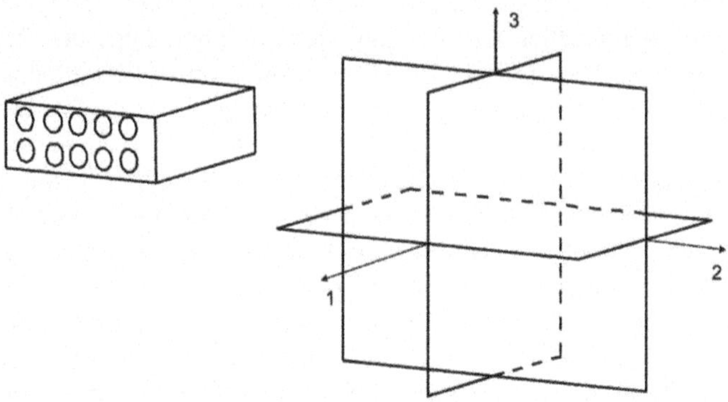

Fig. 6.2 Local directions and planes in the monolayer [2]

Transversely isotropic fiber have axial and transverse elastic moduli E_A, E_T, G_A, G_T and Poisson ratios ν_A, ν_T: Young's modulus along the fiber axis is defined as $E_A = \frac{\sigma_1}{\varepsilon_1}$; transverse direction Young's modulus is defined as $E_T = \frac{\sigma_2}{\varepsilon_2}$; axial shear modulus (constitutes the relation between shear stress and shear strain) is defined as $G_A = \frac{\tau_{12}}{\gamma_{12}} = \frac{\tau_{13}}{\gamma_{13}}$; transverse shear modulus is defined as $G_T = \frac{\tau_{23}}{\gamma_{23}}$; axial Poisson's ratio is $\nu_A = \frac{\varepsilon_2}{\varepsilon_1}$ and, finally, transverse Poisson's ratio is $\nu_T = \frac{\varepsilon_3}{\varepsilon_2}$.

Isotropic matrix is characterized by Young's modulus E_m and Poisson's ratio ν_m. Fiber volume fraction in monolayer is a quotient of the fiber volume and overall volume of the material. For example, in the RVE on Fig. 6.3, fiber volume fraction equals the quotient of two cylinders to the volume of bounding box: $V_f = \frac{\pi (d_f/2)^2}{2a_2a_3}$, where d_f is a fiber diameter.

In the local monopoly coordinate system 123 (Fig. 6.2) A index in fiber properties relates to the indexes 1, 12 and 13, while T index relates to 2, 3 and 23.

Stresses. In the process of elastic body deformation, internal forces resisting the deformation occur. These forces are called internal stresses (direct and shear), which are defined by tensor of 2nd rank—stress tensor. In Fig. 6.4 quantities of the elastic tensor are illustrated. In the following expressions the Voigt contracted notation is used for indication of components of stress tensor: 11 $\Leftrightarrow$ 1, 22 $\Leftrightarrow$ 2, 33 $\Leftrightarrow$ 3, 12 $\Leftrightarrow$ 6, 13 $\Leftrightarrow$ 5, 23 $\Leftrightarrow$ 4, where the indexes in contracted notation are chosen using the two rules: for diagonal terms only single number in index holds (e.g. 11 becomes just 1), for off-diagonal terms the index number in contracted notation is obtained using simple expression: $9 - i - j$ (e.g., 12 becomes 6).

Stiffness tensor. Stiffness tensor constitutes the relation between tensor strains and tensor stresses in elastic materials as follows:

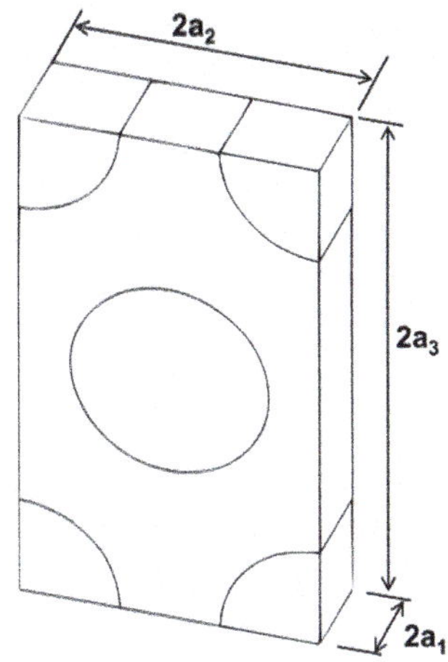

Fig. 6.3 Representative volume element of the composite material with hexagonal pack of fibers [29]

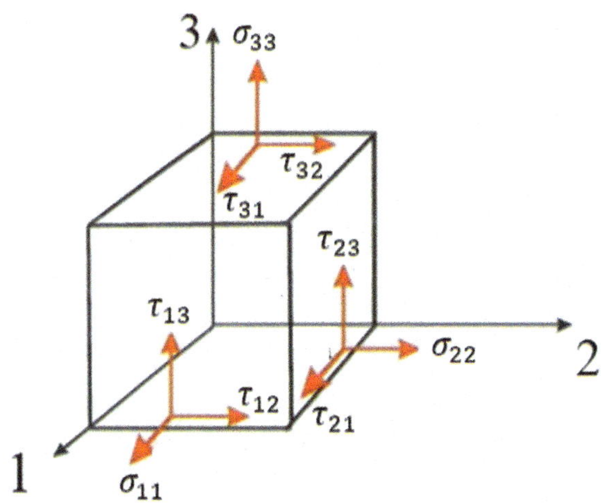

Fig. 6.4 Stress tensor quantities

$$\sigma_{ij} = C_{ijkl}\varepsilon_{kl},$$

where σ_{ij} are components of the stress tensor, ε_{kl} are components of the strain tensor, C_{ijkl} are components of the stiffness tensor.

Output parameters of homogenization procedure would be elastic properties of the transversely isotropic composite monolayer. The notation used refers to directions and planes in Fig. 6.2: transversely isotropic material of the monolayer requires five independent elastic properties for identification: $E_1, E_2, G_{12}, \nu_{12}, \nu_{23}$, while other might be expressed in terms of those five ($G_{13} = G_{12}, \nu_{13} = \nu_{12}$, $G_{23} = \frac{1}{2}E_2/(1 + \nu_{23})$). These five parameters are defined as follows: axial Young's modulus of monolayer $E_1 = \sigma_1/\varepsilon_1$; transverse Young's modulus of monolayer: $E_2 = \sigma_2/\varepsilon_2$; axial Poisson's ratio $\nu_{12} = \frac{\varepsilon_2}{\varepsilon_1}$; transverse Poisson's ratio $\nu_{23} = \frac{\varepsilon_3}{\varepsilon_2}$;

$U[0, 2], R_2 \sim U[0, 2\pi], R_3 \sim U[0.5; 1.5], R_4 \sim U[0, 1]$; U is a uniform distribution on the given interval. Parameter R_1 controls the magnitude of the coordinate increment and decreases along the iteration count, thus localizing search procedure. Parameter R_2 might change the direction of search along the coordinate, parameter R_3 changes the magnitude of significance of leader following process, parameter R_4 controls the evolution law of the solution coordinate.

Gaining-Sharing knowledge method [15] imitates the transformation of the human knowledge. People tend to interact with each other, cooperate, compete, share experience. In the early stages of growth, individual gains knowledge from environment: family, neighborhood, cousins. Later, in adulthood, individual interacts with wider environment, such as working colleagues and social network. At this moment process of gaining knowledge is complemented with the process of sharing knowledge. The example is IT industry, where developers are subdivided into junior, middle and senior roles.

Junior developer is usually a person with basic knowledge of algorithm theory and programming, while this knowledge has negligible background of practice.

Middle developer often possesses practical experience, which is enough to solve the standard and non-standard tasks independently and share the knowledge with junior developers.

Senior developer is an expert in the local field of applied tasks and has significant experience. The developer of this rank can create the architecture of the project, take global solutions and organize the work of developer team.

Nominally, one can identify two phases in the process of gaining and sharing knowledge: junior-middle phase and middle-senior phase, where the professional advancement is happening. Each of the phases has peculiarities, which are imitated in optimization algorithm.

During the first phase (junior-middle), the interaction occurs mainly in the domain of nearest local colleagues, including accidental contacts, while in the second phase (middle-senior) the interchange of knowledge happens between representatives of different groups of developers, and this type of interaction dominates with time.

Both phases are taking place simultaneously, what is more in the beginning of the extremum search the specific operators of the first phase dominate, and with a progress of solution—of the second phase.

In the beginning of solution, uniformly distributed set of initial points NP in the feasible set D is stochastically generated (mimicking number of individuals in population): $I = \{x^j = (x_1^j, x_2^j, \ldots, x_n^j)^T, j = 1, \ldots, NP\} \subset D$, where x^j is a coordinate vector of j individual, NP is a number of individuals in population. All individuals of the population (solutions) are ranked in the ascending order of the corresponding objective function values: $x^1 = x^{best}, \ldots, x^{j-1}, x^j, x^{j+1}, \ldots, x^{NP} = x^{worst}$.

In the first phase (junior-middle) for each of the individuals the nearest neighbors are identified $x^{j,1}, x^{j,2}$ (it is presumed, that the position of the first neighbour $x^{j,1}$ is better than position $x^{j,2}$ of the second one, which depends on the corresponding value of objective function):

- for the solution $x^j, j \neq 1, j \neq NP$, neighbors are $x^{j,1} = x^{j-1}, x^{j,2} = x^{j+1}$;
- for the leader of population $x^1 = x^{best}$ $(j = 1)$ the neighbors are taken from the "right side": $x^{1,1} = x^2, x^{1,2} = x^3$;
- for the worst individual $x^{NP} = x^{worst}$ $(j = NP)$ the neighbors are taken from the "left side": $x^{NP,1} = x^{NP-1}, x^{NP,2} = x^{NP-2}$.

For the process of sharing knowledge a random individual x^r from the rest of population (except for the individual and his two neighbors) is taken.

If $f(x^j) > f(x^r)$, meaning that position of the individual $x^j, j = 1, \ldots, NP$, is better than x^r, the updated position of coordinate i of individual x^j is obtained by:

$$x_i^{j,new} = x_i^j + k_f\left[\left(x_i^{j,1} - x_i^{j,2}\right) + \left(x_i^r - x_i^j\right)\right], i = 1, \ldots, n.$$

If $f(x^j) \leq f(x^r)$, meaning that position of the individual x^j isn't better than x^r, the updated position of coordinate i of individual x^j is obtained by:

$$x_i^{j,new} = x_i^j + k_f\left[\left(x_i^{j,1} - x_i^{j,2}\right) + \left(x_i^j - x_i^r\right)\right], i = 1, \ldots, n,$$

where $k_f \in [0, 1]$ is a knowledge gaining rate.

In the second phase (middle-senior) the population is split into three groups, depending on the objective function value:

- the best individuals group $N^{best} = [b \cdot NP]$ of the first individuals by rank, where $b \in [0; 0.4]$;
- the worst individuals group $N^{worst} = [w \cdot NP]$ of the last individuals by rank, where $w \in [0; 0.4]$;
- the middle individuals group from the rest of the population: $N^{middle} = NP - N^{best} - N^{worst}$.

For each individual x^j from each group the random representative is selected: $x^{r\_best}, x^{r\_worst}, x^{r\_middle}$.

If $f(x^j) > f(x^{r\_middle})$, meaning that position of individual $x^j, j = 1, \ldots, NP$ is better than position of x^{r_middle}, then updated value of coordinate i of individual x^j is obtained by:

$$x_i^{j,new} = x_i^j + k_f\left[\left(x_i^{r\_best} - x_i^{r\_worst}\right) + \left(x_i^{r\_middle} - x_i^j\right)\right], i = 1, \ldots, n.$$

If $f(x^j) \leq f(x^{r\_middle})$, meaning that position of individual $x^j, j = 1, \ldots, NP$ is worse than position of x^{r_middle}, then updated value of coordinate i of individual x^j is obtained by:

$$x_i^{j,new} = x_i^j + k_f\left[\left(x_i^{r\_best} - x_i^{r\_worst}\right) + \left(x_i^j - x_i^{r\_middle}\right)\right], i = 1, \ldots, n,$$

where $k_f \in [0, 1]$ is a knowledge gaining rate.

For each individual $x^j, j = 1, \ldots, NP$, within current iteration k, the sets of coordinate indexes $I^j_{junior-middle}$, updated by the equations for first phase, and $I^j_{middle-senior}$, updated by the equations for second phase, are obtained.

This is achieved by the following procedure:

- number of coordinates $n_{junior-middle}$, updated by the equations for the first phase, is obtained:

$$n_{junior-middle} = \left[n \left(1 - \frac{k}{K} \right)^{\delta} \right],$$

where $\delta > 0$ is a real value (usually assumed in the range $\delta = 0.5;\ 1;\ 1.5;\ 2$), K is a prescribed maximum number of iterations, $[\cdot]$ is an integer part of the value;
- among the elements of the set $\{1, \ldots, n\}$ there is a random set $I^j_{junior-middle}$ of vector x^j coordinate numbers that are subject to change according to the formulas of the first phase. The rest of the coordinate numbers of vector x^j is put in the set $I^j_{middle-senior} = \left\{ \{1, \ldots, n\} \backslash I^j_{junior-middle} \right\}$.

At the beginning of the first and second phases of the search, a random number *rand* is generated using a uniform distribution law on the interval [0, 1]. Next, the *rand* value is compared with the threshold value k_r:

- if $rand > k_r$, then $x_i^{j,new} = x_i^j$ (the coordinate does not change);
- if $rand \le k_r$, then the formulas of the first and second phases are implemented, respectively.

The execution of the algorithm is terminated when maximum prescribed number of iterations K is reached.

Calculation of the objective function is made using the following suggested notation: V_f is a fiber volume fraction; $E_{1(\exp)}, E_{2(\exp)}, \nu_{12(\exp)}, \nu_{23(\exp)}, G_{12(\exp)}$ are a known experimental results; $E_A, E_T, \nu_A, \nu_T, G_A, G_T, E_m, \nu_m$ are an adjusted input parameters; $E_1, E_2, \nu_{12}, \nu_{23}, G_{12}$ are an output parameters. The subscripts denote: *m*—matrix properties, *f*—fiber properties, *A*—axial, *T*—transversal.

6.2.3 Algorithm of Objective Function Evaluation

Step 1. Matrix constants evaluation:

$$\lambda_m = \frac{E_m \nu_m}{(1 + \nu_m)(1 - 2\nu_m)}, \ \mu_m = G_m = \frac{E_m}{2(1 + \nu_m)} \text{ is a Lame's coefficients.}$$

Step 2. Calculation of the coefficients s_3, s_6, s_7, which account for geometry of the fibers and their distribution in RVE:

$$s_3 = 0.49247 - 0.47603V_f - 0.02748V_f^2;$$

$$s_6 = 0.36844 - 0.14944V_f - 0.27152V_f^2;$$

$$s_7 = 0.12346 - 0.32035V_f + 0.23517V_f^2.$$

Step 3. Calculation of the stiffness tensor C'_{ij} components for the case of transversely isotropic material:

$$C' = \begin{pmatrix} C'_{11} & C'_{12} & C'_{13} & 0 & 0 & 0 \\ C'_{12} & C'_{22} & C'_{23} & 0 & 0 & 0 \\ C'_{13} & C'_{23} & C'_{33} & 0 & 0 & 0 \\ 0 & 0 & 0 & C'_{44} & 0 & 0 \\ 0 & 0 & 0 & 0 & C'_{55} & 0 \\ 0 & 0 & 0 & 0 & 0 & C'_{66} \end{pmatrix}$$

3.1. $\Delta = \frac{1-2\nu_A^2 E_T/E_A - 2\nu_T^2 - 2\nu_A^2 \nu_T E_T/E_A}{E_A E_T^2};$

3.2. $C'_{11} = \frac{1-\nu_T^2}{E_T^2 \Delta};$

3.3. $C'_{22} = C'_{33} = \frac{1-\nu_A^2 E_T/E_A}{E_A E_T \Delta};$

3.4. $C'_{12} = C'_{13} = \frac{\nu_A E_T/E_A + \nu_A \nu_T E_T/E_A}{E_T^2 \Delta};$

3.5. $C'_{23} = \frac{\nu_T + \nu_A^2 E_T/E_A}{E_A E_T \Delta};$

3.6. $C'_{44} = G_T = \frac{E_T}{2(1+\nu_T)} = (C'_{22} - C'_{23})/2;$

3.7. $C'_{55} = C'_{66} = G_A.$

Step 4. Calculation of the coefficients $a_1, \ldots, a_4$ (for stiffness tensor terms compact representation):

4.1. $a_1 = 4\mu_m^2 - 2\mu_m C'_{33} + 6\lambda_m \mu_m - 2C'_{11}\mu_m - 2\mu_m C'_{23} + C'_{23}C'_{11} + 4\lambda_m C'_{12} - 2 C'^2_{12} - \lambda_m C'_{33} - 2C'_{11}\lambda_m + C'_{11}C'_{33} - \lambda_m C'_{23};$

4.2. $a_2 = 8\mu_m^3 - \mu_m^2 C'_{33} + 12\mu_m^2 \lambda_m - 4\mu_m^2 C'_{11} - 2\mu_m C'^2_{23} + 4\mu_m \lambda_m C'_{23} + 4\mu_m C'_{11} C'_{33} - 8\mu_m \lambda_m C'_{33} - 4\mu_m C'^2_{12} + 2\mu_m C'^2_{33} - 4\mu_m C'_{11}\lambda_m + 8\mu_m \lambda_m C'_{12} + 2\lambda_m C'_{11} C'_{33} + 4C'_{12}C'_{23}\lambda_m - 4\lambda_m C'_{12}C'_{33} - 2\lambda_m C'_{11}C'_{23} - 2C'_{23}C'^2_{12} + C'^2_{23}C'_{11} + 2C'_{33}C'^2_{12} - C'_{11}C'^2_{33} + \lambda_m C'^2_{33} - \lambda_m C'^2_{23};$

4.3. $a_3 = \frac{4\mu_m^2 + 4\lambda_m\mu_m - 2C'_{11}\mu_m - 2\mu_m C'_{33} - C'_{11}\lambda_m - \lambda_m C'_{33} - C'^2_{12} + C'_{11}C'_{33} + 2\lambda_m C'_{12}}{a_2} - \frac{s_3 - \frac{s_6}{2-2\nu_m}}{\mu_m};$

4.4. $a_4 = -\frac{-2\mu_m C'_{23} + 2\lambda_m \mu_m - \lambda_m C'_{23} - C'_{11}\lambda_m - C'^2_{12} + 2\lambda_m C'_{12} + C'_{11}C'_{23}}{a_2} + \frac{s_7}{\mu_m(2-2\nu_m)}.$

Step 5. Stiffness tensor C^* components calculation for the case of composite with cylindrical fibers and fiber volume fraction V_f using Fourier expansion series:

5.1. $A = \left(\frac{(2\mu_m+2\lambda_m-C'_{33}-C'_{23})(a_3^2-a_4^2)}{a_1} + \frac{2(a_4-a_3)(\lambda_m-C'_{12})^2}{a_1^2}\right)^{-1}.$

5.2. $C^*_{11} = \lambda_m + 2\mu_m - V_f(-a_4^2 + a_3^2)\left(-\frac{(2\mu_m+2\lambda_m-C'_{33}-C'_{23})(a_4^2-a_3^2)}{a_1} + \frac{2(a_4-a_3)(\lambda_m-C'_{12})^2}{a_1^2}\right)^{-1}.$

5.3. $C^*_{12} = \lambda_m + V_f\left(\frac{(\lambda_m-C'_{12})(a_4-a_3)}{a_1}\right)A.$

5.4. $C^*_{22} = \lambda_m + 2\mu_m - V_f\left(\frac{(2\mu_m+2\lambda_m-C'_{33}-C'_{23})a_3}{a_1} + \frac{(\lambda_m-C'_{12})^2}{a_1^2}\right)A.$

5.5. $C^*_{23} = \lambda_m + V_f\left(\frac{(2\mu_m+2\lambda_m-C'_{33}-C'_{23})a_4}{a_1} + \frac{(\lambda_m-C'_{12})^2}{a_1^2}\right)A.$

5.6. $C^*_{44} = \mu_m - V_f\left(\frac{2}{2\mu_m-C'_{22}+C'_{23}} - \left(2s_3 - \frac{4s_7}{2-2\nu_m}\right)\mu_m^{-1}\right)^{-1}.$

5.7. $C^*_{66} = \mu_m - V_f\left((\mu_m - C'_{66})^{-1} - \frac{s_3}{\mu_m}\right)^{-1}.$

Step 6. Resulting stiffness tensor C components calculation:

$$C = \begin{pmatrix} C_{11} & C_{12} & C_{13} & 0 & 0 & 0 \\ C_{12} & C_{22} & C_{23} & 0 & 0 & 0 \\ C_{13} & C_{23} & C_{33} & 0 & 0 & 0 \\ 0 & 0 & 0 & C_{44} & 0 & 0 \\ 0 & 0 & 0 & 0 & C_{55} & 0 \\ 0 & 0 & 0 & 0 & 0 & C_{66} \end{pmatrix}.$$

6.1. $C_{11} = C^*_{11};$
6.2. $C_{12} = C_{13} = C^*_{12};$
6.3. $C_{22} = C_{33} = \frac{3}{4}C^*_{22} + \frac{1}{4}C^*_{23} + \frac{1}{4}C^*_{44};$
6.4. $C_{23} = \frac{1}{4}C^*_{22} + \frac{3}{4}C^*_{23} - \frac{1}{4}C^*_{44};$
6.5. $C_{44} = C^*_{44} = \frac{1}{2}(C_{22} - C_{23});$
6.6. $C_{55} = C_{66} = C^*_{66}.$

Step 7. Resulting compliance tensor S calculation:

$$S = C^{-1} = \begin{pmatrix} S_{11} & S_{12} & S_{13} & 0 & 0 & 0 \\ S_{12} & S_{22} & S_{23} & 0 & 0 & 0 \\ S_{13} & S_{23} & S_{33} & 0 & 0 & 0 \\ 0 & 0 & 0 & S_{44} & 0 & 0 \\ 0 & 0 & 0 & 0 & S_{55} & 0 \\ 0 & 0 & 0 & 0 & 0 & S_{66} \end{pmatrix}$$

(in the following this notation is used S_{11}, S_{12}, S_{22}, S_{23}, S_{44}, S_{55}).

Step 8. Output parameters calculation:

8.1. $E_{1(calc)} = 1/S_{11};$
8.2. $E_{2(calc)} = E_{3(calc)} = 1/S_{22};$

8.3. $\nu_{12(calc)} = \nu_{13(calc)} = -S_{12}/S_{11}$;
8.4. $\nu_{23(calc)} = -S_{23}/S_{22}$;
8.5. $G_{12(calc)} = G_{13(calc)} = 1/S_{55}$;
8.6. $G_{23(calc)} = \frac{E_2}{2(1+\nu_{23})} = 1/S_{44}$.

Step 9. Objective function value is calculated by:

$$\Sigma = \lambda_1\left(\frac{E_{1(calc)} - E_{1(\exp)}}{E_{1(\exp)}}\right)^2 + \lambda_2\left(\frac{E_{2(calc)} - E_{2(\exp)}}{E_{2(\exp)}}\right)^2 + \lambda_3\left(\frac{\nu_{12(calc)} - \nu_{12(\exp)}}{\nu_{12(\exp)}}\right)^2 + \lambda_4\left(\frac{\nu_{23(calc)} - \nu_{23(\exp)}}{\nu_{23(\exp)}}\right)^2 + \lambda_5\left(\frac{G_{12(calc)} - G_{12(\exp)}}{G_{12(\exp)}}\right)^2.$$

6.3 Results

It is assumed, that $E_{1(\exp)}, E_{2(\exp)}, \nu_{12(\exp)}, \nu_{23(\exp)}, G_{12(\exp)}$ parameters are known from experiment and fiber volume fraction V_f is given (Table 6.1).

The constrained feasible set of the unknown parameters values is given in the form of closed intervals $[x_{i\min}, x_{i\max}]$, $i = 1, \ldots, n$: $E_A \in [200{,}000,\ 600{,}000]$ MPa; $E_T \in [8000,\ 30{,}000]$ MPa; $G_A \in [5000,\ 90{,}000]$ MPa; $\nu_A \in [0.15; 0.4]$; $\nu_T \in [0.2; 0.4]$; $E_m \in [3000,\ 10{,}000]$ MPa; $\nu_m \in [0.2; 0.4]$. For a given problem, the solution from [30] is known and taken here for reference (Table 6.2).

It is then assumed, that all of the terms in objective function have the same influence, i.e. $\lambda_1 = \ldots = \lambda_5 = 1$. In view of the different scale of parameters, presented in the expression for objective function, and implementation features of metaheuristic algorithms, normalization is implemented. This way, it is assumed, that all normalized unknown parameters x_i^{norm}, $i = 1, \ldots, n$, have values in the range [0, 1]. The search process is conducted within parallelepiped set of feasible solutions, dimensionality of which is defined by number of independent variables (in

Table 6.1 Composite properties known from experiment

Elastic constants					Fiber volume fraction
E_1 (MPa)	E_2 (MPa)	ν_{12}	ν_{23}	G_{12} (MPa)	V_f
337,682	6918	0.3	0.52	4755	0.62

Table 6.2 Reference constituent properties [30]

Fiber properties					Matrix properties	
E_A (MPa)	E_T (MPa)	G_A (MPa)	ν_A	ν_T	E_m (MPa)	ν_m
517,000	11,158	10,636	0.269	0.306	4629.558	0.363

Table 6.3 Resulting composite constituents properties

Algorithm	Fiber properties				Matrix properties	
	E_A (MPa)	E_T (MPa)	ν_A	ν_T	E_m (MPa)	ν_m
SCA	439,553	8056	0.25	0.3	5048	0.3415
GSKA	489,289	10,386	0.28	0.296	4995	0.3414

this case $n = 7$). The calculation of the objective function values is conducted with parameters returned to initial scales: $x_i = x_{i\,\min} + x_i^{norm}(x_{i\,\max} - x_{i\,\min}),\ i = 1, \ldots, n$.

The computational procedure was conducted using following computer characteristics: Intel Core i5-9400F processor with 2.90 GHz frequency and 16 Gb RAM of DDR3 type. The integrated development environment Microsoft Visual Studio 2022 and C# 7.0 was used for programming and execution of the code.

The results of implementation of the Sine–Cosine method and Gaining-Sharing Knowledge method is summarized in Table 6.3. Following SCA algorithm parameters were taken: maximum iteration count $K = 100$, population size $NP = 50$. Following GSK algorithm parameters were taken: number of individuals in population $NP = 500$; maximum iteration count $K = 1000$; knowledge gaining rate $k_f = 0.4$; threshold value $k_r = 0.9$; best individuals fraction $b = 0.1$, worst individuals fraction $w = 0.1$; modification coefficient for the number of coordinates in junior-middle phase $\delta = 0.5$.

The minimum value of the objective function, achieved within each method is respectively: $\sum_{SCA} = 1.934 \times 10^{-17}$, $\sum_{GSKA} = 1.727 \times 10^{-20}$. Such values confirm that the solutions of the reasonable accuracy were determined.

6.4 Conclusion

Determination of the constituent properties of composite material improves the accuracy of modeling of material at micro scale, which in turn allows for accounting for morphology of PCM and prediction of equivalent elastic properties. Additionally, information about material components properties is important for the design of the new materials. Elastic properties of matrix and fiber can be used for deciding optimum fabrication regimes and for quality control after manufacturing. Meanwhile, most of the micromechanical elastic properties are hard or even impossible to determine experimentally at the state-of-art technology level. Proposed method of the mathematical modeling and parametric optimization in conjunction with periodic microstructure modeling (PMM) technique allow for prediction of the composite constituent's properties with reasonable accuracy.

Acknowledgements The authors wish to thank Eva S. Timofeeva for the help in computation.

References

1. J. Zhu, W. Zhang, L. Xia, Topology optimization in aircraft and aerospace structures design. Arch. Comput. Methods Eng. **4**, 595–622 (2015)
2. E. Barbero, *Introduction to Composite Materials Design*, chap. 4, 3rd edn. (CRC Press, Boca Raton, FL, 2017)
3. M. Paley, J. Aboudi, Micromechanical analysis of composites by the generalized cells model. Mech. Mater. **14**(2), 127–139 (1992)
4. Z. Hashin, S. Shtrikman, A variational approach to the theory of the elastic behaviour of multiphase materials. Mech. Phys. Solids **11**(2), 127–140 (1963)
5. T. Mori, K. Tanaka, Average stress in matrix and average elastic energy of materials with misfitting inclusions. Acta Metall. Mater. **21**(5), 571–574 (1973)
6. R. Hill, The elastic behavior of a crystalline aggregate. Proc. Phys. Soc. Lond. **65**, 349–354 (1952)
7. B. Budiansky, On the elastic moduli of some heterogeneous materials. J. Mech. Phys. Solids **13**(4), 223–227 (1965)
8. J. Eshelby, The determination of the elastic field of an ellipsoidal inclusion and related problems. Proc. R. Soc. A **241**, 376–396 (1957)
9. J. Eshelby, The elastic field outside an ellipsoidal inclusion. Acta Metall. A **252**, 561–569 (1959)
10. S. Nemat-Nasser, M. Hori, *Micromechanics: Overall Properties of Heterogeneous Materials* (North-Holland, Amsterdam, 1993)
11. R. Luciano, E. Barbero, Formulas for the stiffness of composites with periodic microstructure. Int. J. Solids Struct. **31**, 2933–2944 (1994)
12. E. Barbero, R. Luciano, Micromechanical formulas for the relaxation tensor of linear viscoelastic composites with transversely isotropic fibers. Int. J. Solids Struct. **32**, 1859–1872 (1995)
13. C. Floudas, P. Pardalos, *Encyclopedia of Optimization*, 2nd edn. (Springer, New York, NY, USA, 2009)
14. M. Gendreau, *Handbook of Metaheuristics*, 2nd edn. (Springer, New York, NY, USA, 2010)
15. X. Yang, S. Chien, T. Ting, *Bio-Inspired Computation and Optimization*, 1st edn. (Morgan Kaufmann, New York City, NY, USA, 2015)
16. Z. Beheshti, S. Shamsuddin, A review of population-based meta-heuristic algorithms. Int. J. Adv. Soft Comput. Appl. **5**, 1–35 (2013)
17. M. Roni, M. Rana, H. Pota, M. Hasan, M. Hussain, Recent trends in bio-inspired meta-heuristic optimization techniques in control applications for electrical systems: a review. Int. J. Dyn. Control **10**, 999–1011 (2022)
18. A. Panteleev, A. Kolessa, Optimal open-loop control of discrete deterministic systems by application of the perch school metaheuristic optimization algorithm. Algorithms **15**, 157 (2022)
19. A. Kumar, M. Nadeem, H. Banka, Nature inspired optimization algorithms: a comprehensive overview. Evol. Syst. **14**, 141–156 (2023)
20. D. Davendra, I. Zelinka, *Self-organizing Migrating Algorithm. Methodology and Implementation*, in *Studies in Computational Intelligence*, vol. 626, 1st edn. (Springer, New York, NY, USA, 2016)
21. E. Niculina Dragoi, V. Dafinescu, Review of metaheuristics inspired from the animal kingdom. Mathematics **9**, 2335 (2021)
22. A. Tzanetos, I. Fister, G. Dounias, A comprehensive database of nature-inspired algorithms. Data Brief **31**, 105792 (2020)
23. I. Fister Jr., X.-S. Yang, I. Fister, J. Brest, D. Fister, A brief review of nature-inspired algorithms for optimization. arXiv:1307.4186 (2013)
24. M. Karane, A. Panteleev, Multi-agent algorithms for optimizing bundles of trajectories of deterministic systems with incomplete instant feedback. J. Comput. Syst. Sci. Int. **61**(5), 751–775 (2022)

25. A. Panteleev, M. Karane, Application of a novel multi-agent optimization algorithm based on PID controllers in stochastic control problems. Mathematics **11**, 2903 (2023)
26. S. Mirjalili, SCA: a sine cosine algorithm for solving optimization problems. Knowl.-Based Syst. **96**, 120–133 (2016)
27. S. Ramzy, S. Dunia, Chaotic sine-cosine optimization algorithms. Int. J. Soft Comput. **13**, 108–122 (2018)
28. A. Vagdy, A. Hadi, A. Khater, Gaining-sharing knowledge based algorithm for solving optimization problems: a novel nature-inspired algorithm. Int. J. Mach. Learn. Cybern. **11**(5) (2020)
29. E. Barbero, *Finite Element Analysis of Composite Materials Using Abaqus, chap. 6*, 2nd edn. (CRC Press, Boca Raton, FL, 2023)
30. E. Barbero, *Finite Element Analysis of Composite Materials Using Abaqus*, chap. 11, 2nd edn. (CRC Press, Boca Raton, FL, 2023)

Chapter 7
Physical Principles of Wastewater Heat Utilization Using High-Temperature Solar Installations

Gulom Uzakov, Bobir Toshmamatov, and Nasiba Abdurakhmanova

Abstract In world practice, the use of renewable energy sources and heat pump devices in creating modern energy and resource-saving systems for managing traditional fuel and energy resources is increasing. Experimental research, technical measurements, and heat and technical calculation methods were used to assess the efficiency of obtaining technical clean water by utilizing the low-potential heat of wastewater at the "Shurtan Gas-Chemical Complex" enterprice. Analysis of wastewater's physical and thermal properties from "Shurtan Gas-Chemical Complex" enterprice. Empirical equations were obtained that determine the physical properties of wastewater formed from the addition of oily, chemically contaminated, and domestic wastewater generated in the technological processes of "Shurtan Gas-Chemical Complex" enterprice. As a result, it was determined that the density of wastewater is $\rho = 920.2$ kg/m^3, and the coefficient of dynamic viscosity is $\mu = 0.00496$ m^2/s. A technological scheme was developed that allows obtaining technical water by utilizing the heat of wastewater from "Shurtan Gas-Chemical Complex" enterprice using parabolic cylindrical solar concentrator, and equations were obtained that allow calculating its energy efficiency. For the technological process, 6 Eurotrough-150 solar concentrators are connected in series, with a reflector surface area of 408.75 m^2 and a length of 74.25 m. According to the technical parameters of the Eurotrough-150 solar concentrator, experimental studies and calculation results, it was determined that the Eurotrough-150 solar concentrator in the natural climatic conditions of the city of Karshi has an average solar radiation of 700–800 W/m^2, $\eta_{PCSC} = 19.54\%$, when the Re number of wastewater is 10,000, $\eta_{PCSC} = 39.08\%$ when Re = 20,000 and $\eta_{PCSC} = 97\%$ when Re = 50,000.

G. Uzakov (✉) · B. Toshmamatov
Karshi State Technical University, Karshi, Uzbekistan
e-mail: uzoqov66@mail.ru

N. Abdurakhmanova
"Shurtan GChC" LLC, Guzar d, Uzbekistan

© The Author(s), under exclusive license to Springer Nature Switzerland AG 2025

A. Gibadullin et al. (eds.), *Proceedings of the 4th International Scientific and Practical Symposium on Materials Science and Physics*, Springer Proceedings in Physics 318,
https://doi.org/10.1007/978-3-031-94805-3_7

7.1 Introduction

As a result of the significant reduction in the amount of natural fuel and energy resources in the world and the increase in emissions of toxic emissions into the environment, the conservation of traditional energy sources and their use in energy-resource-efficient devices is very urgent [1]. Today, the problem of energy resources has become global in all countries, and the restructuring of the energy base, the use of environmentally friendly, renewable, and alternative energy sources, and low-potential secondary energy resources for heat recovery processes is rapidly developing [2, 3].

Secondary energy resources (SER) in industrial and manufacturing enterprises are low-potential SER, such as the heat of combustion products generated during the combustion of fuel and energy resources, the heat of wastewater and air generated after the cooling process in technological devices, the heat of ventilation air, the secondary heat of water vapor in technological processes, the heat of chemically contaminated wastewater and sewage, etc. [4].

7.1.1 Statement of the Problem

In oil and gas processing plants, a large amount of low-potential heat is released into the environment with wastewater and causes environmental damage.

Technical clean water used in oil and gas processing plants is contaminated with various hydrocarbon compounds (oils, petroleum oils), various types of salts, coarse, colloid, and particles, as well as mechanical impurities, and in most cases such chemically contaminated water is technically treated and discharged as wastewater [5, 6].

The consumption of water used during technological processes at enterprises, its chemical composition, quality indicators, and the composition of the resulting chemically contaminated water depend on the current state of production technologies, the type of product being produced, the level of technical equipment, and automation of the enterprise.

Wastewater is rarely used as a coolant, and the main focus in the design of energy-efficient systems is on data on the temperature in wastewater collection collectors, water flow rate, and physicochemical properties [7, 8].

7.2 Materials and Methods

The wastewater of the "Shurtan GChC" LLC enterprise is a liquid mixture of waste generated during the production and domestic activities of the enterprise. It represents water containing a mixture of dissolved and undissolved liquid, solid, and gaseous substances.

All methods used in wastewater treatment are classified into 3 types according to their essence and content, namely:

- Mechanical—settling, centrifugation, and filtration.
- Chemical—combining various wastewaters and carrying out chemical reactions, for example, neutralization reactions, to remove harmful substances.
- Biological—with microorganisms, mycobacteria, and aquatic plants.

At the "Shurtan GChC" LLC enterprise, oily wastewater is 10–36 m^3 per hour, 864 m^3 per day, chemical wastewater is 18 m^3 per hour, 432 m^3 per day, domestic wastewater is 40 m^3 per hour, 960 m^3 per day, and their composition, physical and chemical properties are presented in Tables 7.1 and 7.2.

Chemically contaminated wastewater consists of various organic and inorganic compounds and must be technically treated depending on technological properties before being reused and discharged as wastewater.

Mechanical methods of wastewater treatment.

Table 7.1 "Shurtan GChC" LLC results of wastewater analysis for 11.06.2024 [4]

Component name (requirements)	Identified component cost (requirements)	
	According to the regulatory document	Fact
I	II	III
pH	6.5–8.5	8.16
Suspended solids, mg/l	30	30
Ammonium nitrogen, mg/l	2.0	1.8
Nitrite ion, mg/l	3.3	0.066
Nitrate ion, mg/l	45	8.3
Chlorides, mg/l	350	329
Petroleum product, mg/l	0.3	0.22
Phosphate ion, mg/l	1.0	0.36
Sulfate ion, mg/l	500	472
Dry residue, mg/l	1000	976
Ferric ion (+3), mg/l	0.3	0.12
Dissolved oxygen, mg/l	4–6	5.42
BOD, mg/l	6.0	–
CDP, mg/l	40	36.0

Table 7.2 "Shurtan GChC" LLC results of wastewater analysis for 11.12.2024

Date	Sampling point	Standard	pH	Dissolved oxygen	Mg/l						
					Suspended articles	Dry residue	Petroleum product	Iron	Ammonium nitrogen	CPC	Residual chlorine
11.12.2024	Oily wastewater Outlet from GA-6401/S pump	Standard	6.5–8.15		63	1747	1.5	0.13	16.4	134.4	
		Fact	6.71		38	976	1.1	0.3	4.6	84	
	Chemical wastewater	Standard	7.5–8.9		44	1982		0.15	24		
		Fact	8.47		42	1966		0.14	6.4		
	Domestic wastewater	Standard	6.5–8.5		31	997		0.11	15.5		
		Fact	7.14		18	780		0.1	7.42		
	Contact chlorination pool outlet	Standard	6.5–8.5	4.0	23.14	1500	0.3	0.3	2.0		0.3–0.5
		Fact	7.38	3.8	20	1850	0.24	0.25	2.88		0.33

Mechanical wastewater treatment is used to remove undissolved mineral and organic impurities from the water being treated.

The implementation of mechanical treatment usually involves the preparation of industrial wastewater to achieve a high level of purification using one of the physicochemical, chemical and biological, as well as thermal methods.

Such treatment provides for the removal of suspended solids from wastewater by 90–95% and a reduction in organic pollution (total BOD) by 20–25%.

More complete clarification of wastewater, the process of cleaning it from suspended particles, is carried out by filtering, that is, passing water through various granular materials (quartz sand, granite gravel, slag formed in cast iron foundries, etc.).

7.2.1 Calculation Methods

Taking into account the thermal and physical properties of oily, chemical and domestic wastewater generated in the technological processes of "Shurtan GChC" LLC, we express the movement of wastewater based on the mass conservation equation as follows [4]:

$$\frac{\partial \rho}{\partial t} + \nabla \cdot (\rho \cdot v) = 0 \tag{7.1}$$

where, ρ—wastewater density, $\frac{kg}{m^3}$; v—velocity vector, $\frac{m}{sek}$.

Based on the Novier-Stoke equation, we express it as follows:

$$\rho(\frac{\partial v}{\partial t} + (\nabla \cdot v)v = -\nabla P + \nabla \cdot (\mu \nabla v) + F \tag{7.2}$$

where, P—pressure, Pa; μ—coefficient of dynamic viscosity of wastewater, $\frac{m^2}{sek}$; F—external forces.

We apply the advection–diffusion equation for each component of wastewater (petroleum oils, chemicals, organics):

$$\frac{\partial \mathrm{C_i}}{\partial \mathrm{t}} + \nabla \cdot (\mathrm{C_i} \cdot \mathrm{v}) = \mathrm{D_i} \cdot \nabla^2 \cdot \mathrm{r_i} + \mathrm{R_i}, \tag{7.3}$$

where, r_i—i-concentration of the component; D_i—diffusion coefficient; R_i—source of the substance (substance deposition).

We write the temperature equation of wastewater as follows:

$$\rho C_p \frac{\partial T}{\partial t} + \rho C_p v \cdot \nabla T = \lambda \cdot \nabla^2 T + Q, \tag{7.4}$$

T—wastewater temperature, K; C_p—specific heat capacity of wastewater, $\frac{Dj}{(kg \cdot K)}$; λ—thermal conductivity coefficient, $\frac{W}{(m \cdot K)}$; Q—thermal power of wastewater, W.

The thermal and physical properties of wastewater are calculated by the volume fractions of the components that make up the wastewater. Therefore, we calculate the density of wastewater based on the following equation:

$$\rho = \sum_i r_i \cdot \rho_i \tag{7.5}$$

where, $r_i = \frac{V_i}{\sum_i V_i}$—volume fraction of the component, %.

We calculate the dynamic viscosity coefficient of wastewater based on the following equation:

$$\mu = \sum_i r_i \cdot \mu_i \tag{7.6}$$

μ_i—dynamic viscosity coefficient of the component, $\frac{m^2}{sek}$.

We calculate the specific heat capacity of wastewater based on the following equation:

$$C_p = \sum_i r_i \cdot C_{pi} \tag{7.7}$$

We calculate the thermal conductivity of wastewater based on the following equation (Table 7.3):

$$\lambda = \sum_i r_i \cdot \lambda_i \tag{7.8}$$

We calculate the total wastewater consumption using the following equation [5]:

$$G_{ww} = G_{o.w.} + G_{ch.w.} + G_{d.w.} \tag{7.9}$$

We find the average pollution concentration of wastewater.

$$G_{ww} C_{ww} = G_{o.w.} C_{o.w.} + G_{ch.w.} C_{ch.w.} + G_{d.w.} C_{d.w.} \tag{7.10}$$

We write the equation in differential form, taking into account the change in concentration with time:

$$\frac{d(C_{ww.} V)}{dt} = G_{o.w.} \cdot C_{o.w.} + G_{ch.w.} \cdot C_{ch.w.} + G_{d.w.} \cdot C_{d.w.} - G_{out.} \cdot C_{ww.} \tag{7.11}$$

where, $C_{o.w.}$, $C_{o.w.}$, $C_{d.w.}$—concentration of pollutants in wastewater, mg/l; V—working volume of the wastewater treatment plant PCSC, m^3.

Table 7.3 The results of calculations of the thermal and physical parameters of wastewater from "Shurtan GChC" LLC

Water type	Consumption, G, m^3/h	Consumption, G, m^3/day	Density, ρ kg/m^3	Volume fraction of the component, r_i, %	Dynamic viscosity coefficient, μ, $\frac{m^2}{sek}$
Oily waste water, $G_{o.w.}$	10–36	864	800	0.36	0.01
Chemical waste water, $G_{ch.w.}$	18	432	1000	0.18	0.002
Domestic waste water, $G_{d.w.}$	40	960	990	0.46	0.001
Mixed waste water, G_{ww}	68–94	2256	920.2	1.0	0.00496

The temperature of the wastewater mixture is determined by the following formula:

$$G_{ww}T_{ww} = G_{o.w.} \cdot T_{o.w.} + G_{ch.w.} \cdot T_{ch.w.} + G_{d.w.} \cdot T_{d.w.} \tag{7.12}$$

We write the equation in differential form, taking into account the change in temperature with time:

$$\begin{aligned}\frac{d(T_{ww} \cdot \rho \cdot V \cdot C_p)}{dt} &= G_{o.w.} \cdot T_{o.w.} \cdot \rho \cdot C_p + G_{ch.o.} \cdot T_{ch.o.} \cdot \rho \cdot C_p \\ &\quad + G_{d.o.} \cdot T_{d.o.} \cdot \rho \cdot C_p - G_{out} \cdot T_{ww} \cdot \rho \cdot C_p\end{aligned} \tag{7.13}$$

where, C_p—specific heat capacity of wastewater, $\frac{Dj}{(kg \cdot K)}$.

Dividing both sides by $\rho \cdot V \cdot C_p$, we get the following simplified equation

$$\frac{d(T_{ww})}{dt} = \frac{1}{V}(G_{o.w.} \cdot T_{o.w.} + G_{ch.w.} \cdot T_{ch.w.} + G_{d.w.} \cdot T_{d.w.} - G_{out.} \cdot T_{ww.}) \tag{7.14}$$

In practice, the utilization of low-potential heat of wastewater is carried out using various methods. The utilization of wastewater heat in enterprises based on PCSC, that is, its reuse as SER, is of significant scientific and technical importance. The most important feature of the use of this type of equipment is its multifunctionality of low-potential energy. This allows you to optimize the use of fuel and energy resources.

Our country also has various sources of low-potential secondary energy resources, and currently, the utilization of SERs has not been effectively used in power supply systems. In particular, the annual amount of secondary energy sources (wastewater) of "Shurtan Gas-Chemical Complex" LLC is 0.9–1.0 million tons, the hourly consumption is 115 m^3/h, and the average temperature is 20–35 °C.

Utilization of secondary heat of low-potential wastewater using PCSC and obtaining technically purified water is one of the main methods of energy resource saving.

Analytical, heat-technological, heliotechnical, and thermodynamic calculation methods were used to determine the efficiency of utilizing secondary heat of low-potential wastewater generated at the "Shurtan GChC" LLC enterprise and the possibility of obtaining technical water. Based on the conducted experimental studies, measurement, and control works, the main sources of SER of the "Shurtan GChC" LLC enterprise and their heat-technical parameters are presented in Table 7.4.

From the data in Table 7.4, it can be seen that the technical potential of the wastewater of the "Shurtan GChC" LLC enterprise is 3346 kW, and by utilizing this high-power SER based on a parabolic cylindrical solar concentrator, it is possible to achieve energy and resource savings by using technically purified water for the needs of the enterprise and autonomous energy supply systems of the enterprise's facilities.

The proposed scheme of wastewater heat utilization of the "Shurtan GChC" LLC enterprise based on a parabolic cylindrical solar concentrator, shown in Fig. 7.1, allows obtaining technically clean water for the needs of the enterprise by increasing the temperature of wastewater to 150–450 °C (see Fig. 7.1).

Table 7.4 Main thermal and technical parameters of wastewater of the enterprise "Shurtan GChC" LLC

Wastewater (SER) parameters	Marking	Unit of measurement	Quantity
Consumption	G_h	m^3/h	115
Annual quantity	G_y	$m^3/year$	900,000–1 000,000
Pressure	p	MPa	0.05–0.25
Temperature	t	°C	22.4–35
Heat capacity	Q	kW	3346

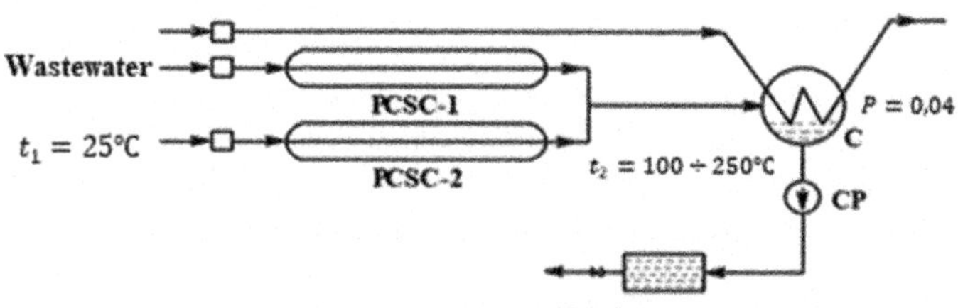

Fig. 7.1 Scheme of wastewater heat utilization based on a parabolic-cylindrical solar concentrator at the "Shurtan GChC" LLC enterprise

F_1, F_2, F_3—filters, PCSC—parabolic cylindrical solar concentrator, C—condenser (heat exchanger), CP—condensate pump.

The solar concentrator reflects the energy of solar radiation and concentrates it on the surface of the absorber tube located in the focal line of the parabolic cylindrical reflector [9, 10]. Due to the geometric shape of the solar concentrator, the parabolic cylindrical reflector collects the direct solar radiation falling on the surface into the absorber tube passing along the focal line. To reflect solar radiation, films made of glass, aluminum or silver are placed on the surface of the parabolic cylindrical surface. The absorber tube is one of the main elements of the solar concentrator, and the efficiency of the device largely depends on it.

To calculate the energy parameters of the PCSC, we use the following calculation methods.

The date (month and day) and time of the design point are selected; solar energy resources are determined for the given coordinates (date and time).

The angle of incidence of direct solar radiation is determined:

The angle of declination (δ) of the sun for a given day n is determined by the following formula [12]:

$$\delta = 23{,}45\sin\left(360\frac{284 + n}{365}\right) \tag{7.15}$$

The hour angle ω of the sun at point A (φ° and ψ°) at a given time of day local time is calculated by the following formula:

$$\omega = \frac{15^\circ}{hour}(12 - t_h) \tag{7.16}$$

where, t_h—sunny time.

Radiation power on the active surface.

Taking into account the amount of direct solar radiation, the solar energy power on the active surface of the PCSC is determined by the following formula [12]:

$$Q_{PCSC} = F_{PCSC} \cdot q^{\Sigma} \cdot cos(\theta) \tag{7.17}$$

F_{PCSC}—PCSC reflector surface area, m^2; q^{Σ}—amount of solar radiation, W/m^2; θ—angle of incidence, °.

The energy efficiency of a PCSC, taking into account all losses in the PCSC (optical, geometric, and thermal), can be written in the following form [13]:

$$\eta_{PCSC} = \eta_{opt,0^\circ} \cdot K(\theta) \cdot t_{h.s.} \tag{7.18}$$

or,

$$\eta_{PCSC} = \frac{Q_{PCSC}^{usef}}{Q_{PCSC}} \tag{7.19}$$

For the case where the PCSC incidence is zero (θ = 0°), the resulting optical losses are expressed as:

$$\eta_{opt,0^\circ} = \partial \cdot \tau \cdot \alpha \cdot \gamma \tag{7.20}$$

For the case where the PCSC incidence angle is greater than zero (θ > 0°), the resulting geometric and optical losses are determined according to the following expression:

$$K = f(\theta) \tag{7.21}$$

The heat losses of the PCSC are determined according to the following expression:

$$Q_{h.l.}^{PCSC} = \Delta q_{h.l..} \cdot L_{PCSC} \tag{7.22}$$

$\Delta q_{h.l..}$—is the specific heat loss (W/m), which is calculated as follows:

$$\Delta q_{h.l.} = \left[\left(0.00036 \cdot \Delta T^2 + 0.2029 \cdot \Delta T + 24.899\right) \cdot \left(\frac{q^\Sigma}{900}\right) \cdot \cos(\theta) \right] + 0.00154 \cdot \Delta T^2 + 0.2021 \cdot \Delta T - 24.899 \tag{7.23}$$

where, L_{PCSC}-PCSC is the length, m; ΔT-PCSC is the difference between the maximum temperature of the wastewater in the absorber pipe and the ambient temperature, K; $\cos(\theta)$-cosine of the angle of incidence, °.

The useful thermal capacity of PCSC depends on solar radiation and the energy parameters of PCSC and is calculated based on the following equation [11].

$$Q_{PCSC}^{usef} = Q_{PCSC} \cdot \eta_{PCSC} \tag{7.24}$$

or

$$Q_{PCSC}^{usef} = F_{PCSC} \cdot q^\Sigma \cdot \cos(\theta) \cdot \eta_{opt,0^\circ} \cdot K(\theta) \cdot t_{h.l.} \cdot k_{dust} \tag{7.25}$$

k_{dust}—dusting coefficient of the reflector surface of the PCSC, ((0 < k_{dust} < 1).

If we replace the expression for thermal efficiency in the equation for useful thermal power with the expression for heat losses, then the equation for calculating the useful thermal power of a PCSC will look like this:

$$Q_{PCSC}^{usef} = F_{PCSC} \cdot q^\Sigma \cdot cos(\theta) \cdot \eta_{opt,0^\circ} \cdot K(\theta) \cdot k_{dust} - Q_{h.l.}^{PCSC} \tag{7.26}$$

Since the daily wastewater consumption of "Shurtan GChC" LLC is high, the number of PCSCs required in the process of technical water extraction using PCSCs is also greater. Therefore, the serially connected PCSC is calculated based on the following equation:

$$N_{PCSC} = \frac{\Delta T_T}{\Delta T_C} \tag{7.27}$$

ΔT_T—maximum temperature of wastewater in the focal area of the PCSC, K: ΔT_C—temperature of wastewater in one PCSC, K.

It is known from the operating mode of the PCSC that the flow regime of the wastewater inside the absorber tube is characterized by the Reynolds number, which prevents the tube from deforming and the PCSC from being damaged when it is overheated [13].

For different Re, the wastewater velocity can be calculated by the following formula:

$$v = \frac{Re \cdot \mu}{\rho \cdot d} \tag{7.28}$$

Here, d is the inner diameter of the absorber pipe, m; ρ is the density of the wastewater, kg/m^3; μ is the coefficient of dynamic viscosity of the wastewater, m^2/s.

After determining the wastewater velocity, the mass flow rate q_m can be determined based on the cross-sectional area S of the PCSC absorber pipe and the density ρ of the liquid:

$$q_m = v \cdot S \cdot \rho \tag{7.29}$$

We also determine the useful thermal capacity of the PCSC through the enthalpies of the PCSC wastewater:

$$Q_{PCSC}^{usef} = q_m \cdot \Delta h = q_m \cdot (h_{out} - h_{in}) = q_m \cdot \int_{T_i}^{T_0} C_p dT \tag{7.30}$$

h_{in}, h_{out}—input and otput enthalpy of wastewater at the inlet to the PCSC, Dj/kg; C_p—specific heat capacity of wastewater, Dj/(kg °C); T_i, T_0—wastewater inlet and outlet temperatures of the PCSC, K.

To obtain technical water through the utilization of "Shurtan GChC" LLC wastewater, a Eurotrough-150 PCSC was selected, its main characteristics are given in Table 7.5 and Fig. 7.2, and a technological scheme of the PCSC utilization process of "Shurtan GChC" LLC wastewater is given.

The calculation results presented in Figs. 7.3, 7.4 and 7.5. For example, Fig. 7.3 shows the results of the dependence of Q_{PCSC}^{usef} on the Reynolds number, that is, as the Reynolds number of the wastewater flowing through the absorber tube of the PCSC changes, the Q_{PCSC}^{usef} also changes, Fig. 7.4 shows the results of the dependence of PCSC thermal power on solar radiation and Fig. 7.5 shows the results of the effect of the useful energy of PCSC on the useful work of the device.

Table 7.5 Main characteristics of PCSC of the Eurotrough-150 brand

Parameter name	Marking	Measurement	Quantity
The width of the parabola	–	m	5.77
PCSC total length	L_{PCSC}	m	74.25
Total number of PCSC	N_{PCSC}	dona	6
The outer diameter of the metal absorber tube	D	m	0.07
The inner diameter of the metal absorber tube	d	m	0.065
PCSC reflector surface	F_{PCSC}	m^2	408.75
Nominal reflective power of a parabolic cylindrical reflector	ρ	–	0.94
Glass lid permeability	τ	–	0.965
Absorption capacity of selective surface	α	–	0.955
Interception coefficient	γ	–	0.90
Maximum optical efficiency	ρ, τ, α, γ	–	0.779

1—"Shurtan GChC" LLC wastewater pond, 2—Eurotrough-150 PCSC, 3—condenser, 4, 6—filter, 5—pump, 7—residual mixture tank, 8—condensate water (technical clean water)

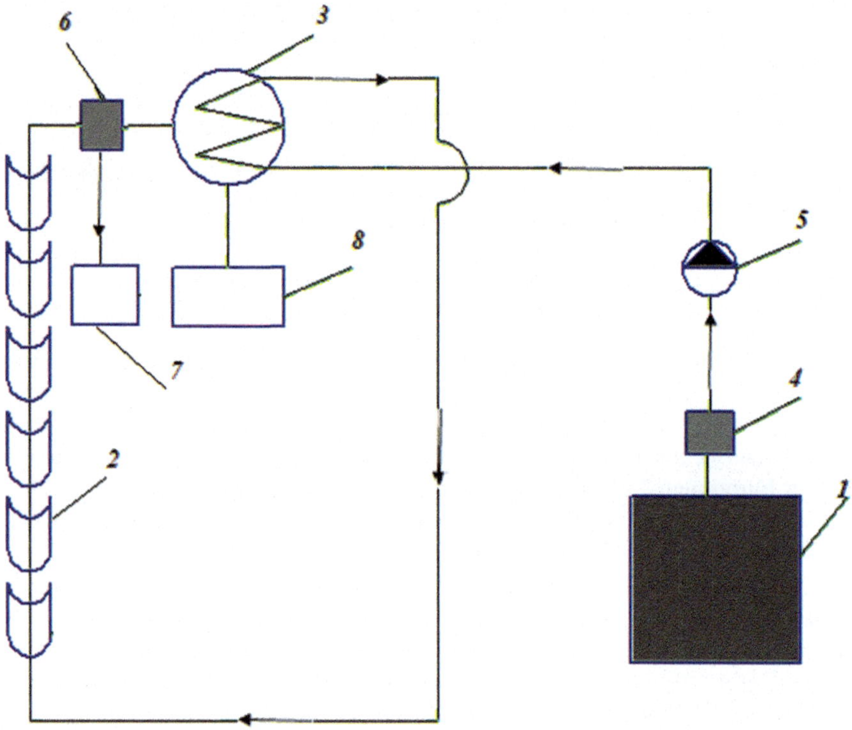

Fig. 7.2 Technological scheme of the process of wastewater disposal of "Shurtan GChC" LLC by PCSC

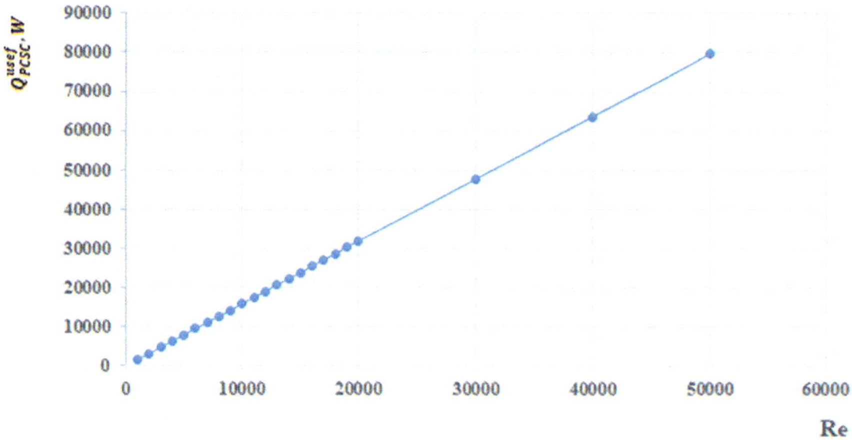

Fig. 7.3 Diagram of the dependence of the useful thermal power of the PCSC on the Re number

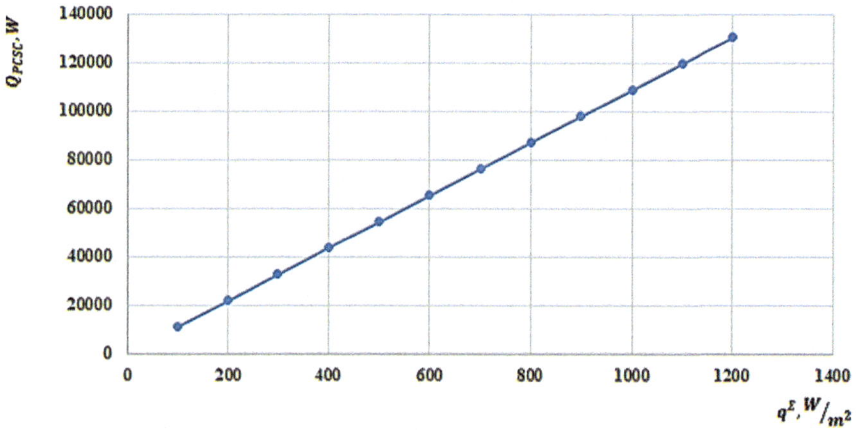

Fig. 7.4 Diagram of the dependence of the thermal power of the PCSC on changes in the amount of solar radiation

7.3 Conclusions

The temperature at the PCSC outlet should be slightly lower than the maximum operating temperature (450 °C) that the oil mixture in the wastewater can reach. In this case, the temperature is set to 430–440 °C.

Rainwater from the complex's technological devices to the oily wastewater treatment device 6401 of the wastewater treatment plant receives 864 m^3 of oily wastewater per hour from 10 to 36 m^3 per hour.

The flue gas is treated using the Oily Wastewater Treatment Plant (Unit 6401) of the chemical complex. The treated wastewater is directed to the PCSC.

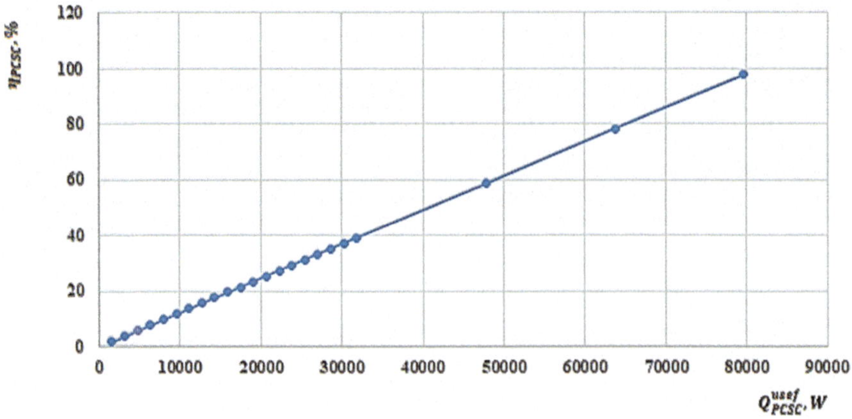

Fig. 7.5 Energy efficiency index of the PCSC

The calculation results were calculated based on the thermal and physical parameters of "Shurtan GChC" LLC wastewater and the main characteristics of the Eurotrough-150 PCSC given in Table 7.3, and the calculation results are presented in Figs. 7.3, 7.4 and 7.5.

As a result, it was determined that the density of the wastewater of the enterprise "Shurtan GChC" LLC is $\rho = 920.2$ kg/m^3, and the dynamic viscosity coefficient $\mu = 0.00496$ m^2/s.

A scheme for utilizing the heat of wastewater at the "Shurtan GChC" LLC enterprise based on a parabolic cylindrical solar concentrator was developed, and a Eurotrough-150 PCSC was selected, and its technical parameters were presented.

A technological scheme for utilizing the heat of wastewater at the "Shurtan GChC" LLC enterprise using PCSC to obtain technical water was developed, and equations were presented that allow calculating its energy efficiency.

6 Eurotrough-150 PCSCs are connected in series to utilize the heat of wastewater at the "Shurtan GChC" LLC enterprise, which allows obtaining technical clean water by increasing the temperature of wastewater to 150–450 °C.

According to the technical parameters of the Eurotrough-150 PCSC, experimental studies and calculation results, it was determined that the Eurotrough-150 PCSC in the natural climatic conditions of the city of Karshi, when the average solar radiation is 700–800 W/m^2, the wastewater Re number is 10,000, $\eta_{PCSC} = 19.54\%$,, when Re = 20,000, $\eta_{PCSC} = 39.08\%$, and when Re = 50,000, $\eta_{PCSC} = 97\%$.

According to the technological scheme, the average temperature of the wastewater of the "Shurtan GChC" LLC enterprise is 20–35 °C, and as a result of the conversion of steam into condensate in the condenser, its temperature rises to 70–85 °C and is directed to the PCSC. As a result, this affects the energy efficiency of the PCSC.

The technical clean water obtained as a result of the utilization of wastewater of the "Shurtan GChC" LLC enterprise using the Eurotrough-150 brand PCSC will be used for technological processes at the enterprise.

References

1. G. Uzakov, S. Khamraev, S. Khuzhakulov, Rural house heat supply system based on solar energy. IOP Conf. Ser.: Mater. Sci. Eng. **1030**(1), 012167 (2021)
2. R. Mehdi, S.R. Naqvi, A.A. Khan, A.A. Mirani, Optimization of olive oil extraction from olive pomace using solvent extraction and response surface methodology analysis of oil yield. Fuel **348**, 128633 (2023)
3. C. Qi, R. Wang, S. Jia, J. Chen, Y. Li, J. Zhang, G. Li, W. Luo, Biochar amendment to advance contaminant removal in anaerobic digestion of organic solid wastes: a review. Biores. Technol. **341**, 125827 (2021)
4. N.K. Abduraxmanova, Efficiency of waste water heat utilization at the "Shurtan Gas Chemical Complex" LLC. Innov. Technol. **2**(54), 29–37 (2024)
5. G. Uzakov et al., Methodology for calculating the potential of secondary thermal energy resources at the Shurtan gas chemical complex. Innov. Technol. **2**(54), 54–61 (2024)
6. K.K.I. Al-Chlaihawi, K. Al-Farhany, A.T. Al-Rubaye, Experimental investigation of solar thermal collector on the open parabolic trough. J. Eng. Appl. Sci. **13**(16), 6873–6878 (2018)
7. S.F. Moosavian, D. Borzuei, A. Ahmadi, Energy, exergy, environmental and economic analysis of the parabolic solar collector with life cycle assessment for different climate conditions. Renew. Energy **165**, 301–320 (2021)
8. M.M. Gubanov, Features of the German legislation in the field of energy saving and application of renewables. Ind. Power **1**, 54–61 (2013)
9. M.A. Islam, F.M. Aldaihani, Justification for adopting qualitative research method, research approaches, sampling strategy, sample size, interview method, saturation, and data analysis. J. Int. Bus. Manag. **5**(1), 01–11 (2022)
10. J. Kudratov et al., Justification of heat-technical parameters of hybrid heat and hot water supply system. Eur. Int. J. Multidiscip. Res. Manag. Stud. **3**, 35–45 (2023)
11. J. Duffy, *Fundamentals of Solar Thermal Energy* (Publishing House "Intellect", 2013)
12. A. Herez, H.E. Hage, T. Lemenand, M. Ramadan, M. Khaled, Review on photovoltaic/thermal hybrid solar collectors: classifications, applications and new systems. Sol. Energy **207**, 1321–1347 (2020)
13. S.S. Kutateladze, *Fundamentals of the Theory of Heat Transfer* (Atomizdat, 1979)

Chapter 8
Local Sodium Carboxymethyl Cellulose: Physicochemical Characteristics and Their Significance

Ummatjon A. Asrorov, Sabitjan Y. Inagamov, Erkin B. Khujanov, Furqat K. Turotov, and Hasan N. Bozorov

Abstract This article presents a description of sodium carboxymethyl cellulose (Na-CMC) polymer produced from cellulose and analyzes its physicochemical properties. Na-CMC is extensively used in industry due to its high efficiency and ability to ensure durability. The research examines the rheological properties of the polymer solution, including viscosity and the loss modulus, as well as the crystalline and amorphous phases present in the material. Samples were prepared at a concentration of 8% and tested at room temperature (25 °C) under specially established conditions using the Anton Paar MCR 92 device.

8.1 Introduction

Cellulose is a naturally occurring polymer derived from plants, chemically highly organized and durable. It primarily constitutes the cell walls of plants, serving as one of their main components. Cellulose's biological strength, renewability, and environmental purity make it widely used in industry. Its components are separated by removing lignin, hemicellulose, and other impurities from plant raw materials (Fig. 8.1) [1, 14].

Primary Sources of Cellulose

1. **Cotton Fiber**. Cotton fiber is the purest source of cellulose, containing up to 99% pure cellulose. It is a high-quality fiber widely used in industries ranging from

U. A. Asrorov (✉) · F. K. Turotov
National University of Uzbekistan Named After Mirzo Ulugbek, Tashkent, Uzbekistan
e-mail: ummat.asrorov1995@gmail.com

U. A. Asrorov · E. B. Khujanov · H. N. Bozorov
Tashkent State Pedagogical University Named After Nizami, Tashkent, Uzbekistan

S. Y. Inagamov
Tashkent Pharmaceutical Institute, Tashkent, Uzbekistan

© The Author(s), under exclusive license to Springer Nature Switzerland AG 2025
A. Gibadullin et al. (eds.), *Proceedings of the 4th International Scientific and Practical Symposium on Materials Science and Physics*, Springer Proceedings in Physics 318,
https://doi.org/10.1007/978-3-031-94805-3_8

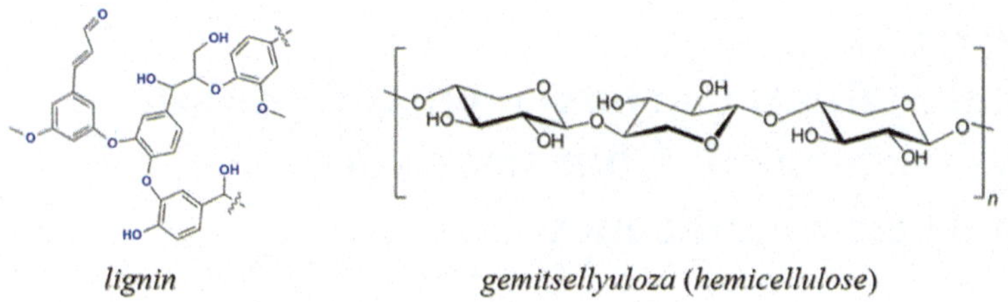

Fig. 8.1 Chemical formula lignin (**a**) and hemicellulose (**b**)

textiles to pharmaceuticals. Cellulose derived from cotton is primarily used to produce sterile medical materials, paper, and other specialized products [1, 2, 14].

2. **Wood**. Wood contains 40–50% cellulose and is considered one of the primary sources. The extraction of cellulose from wood requires the separation of mixtures such as lignin and hemicellulose. Cellulose obtained from wood fibers is often used in paper production, biofuel preparation, and bioplastics [14].
3. **Bamboo and Other Plants**. Bamboo and other fast-growing plants are recognized as economically viable sources of cellulose. Bamboo fibers are rich in cellulose and are used to produce environmentally clean and renewable materials. Cellulose from bamboo is actively used in the textile and food industries (Fig. 8.2) [14].

Cellulose is purified, and some of its hydroxyl groups are substituted with carboxymethyl groups to produce sodium carboxymethyl cellulose [14].

Sodium carboxymethyl cellulose (Na-CMC) is a semi-flexible and negatively charged polyelectrolyte widely used in the chemical industry. Due to its properties, it plays a significant role in the production of various industrial products. With

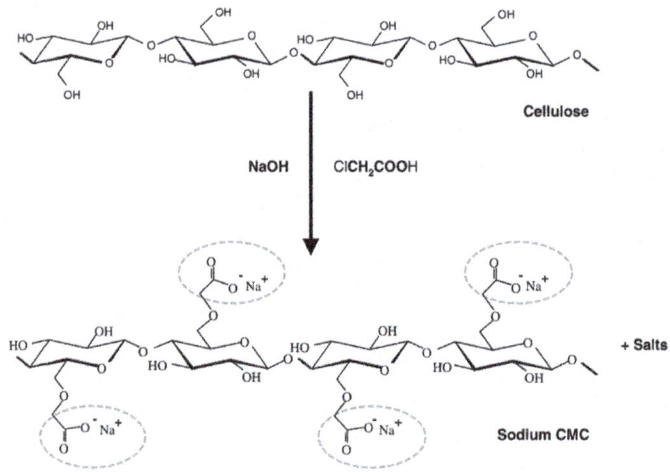

Fig. 8.2 Reaction to form a molecule of Na-CMC from cellulose [14]

its high efficiency and durability, Na-CMC generated a market value of 1.2 billion USD in 2016 and is expected to reach 1.8 billion USD by 2025. This growth trend highlights the importance of Na-CMC in industry and indicates the expanding range of its applications. Na-CMC is primarily used in the composition of various products in the food and beverage, cosmetics, pharmaceutical, and cleaning agents industries. Products containing Na-CMC consist of various components, including solid particles (such as silicon dioxide), polymers, fragrant and flavoring oils, surfactants, salts, or proteins. Creating these complex systems and deeply understanding the interactions among components is essential for defining the efficiency and shelf life of the product. As a biopolymer, Na-CMC exhibits high polydispersity in molecular chain length (with M_w/M_n values typically ranging from 1.7 to 4.5). It is also extensively used in the oil and gas industry. As a thickening component in drilling fluids, Na-CMC plays a crucial role in enhancing process efficiency. The properties of Na-CMC, dependent on the molecular chain length, the amount of carboxymethyl groups, and their distribution along the chain, expand its application possibilities in various fields. Na-CMC products on the market typically contain 0.4–1.5 carboxymethyl groups, a measure known as the degree of substitution (DS), which provides the polymer's water solubility (Tables 8.1 and 8.2) [3–14].

The molecular mass of sodium carboxymethyl cellulose changes depending on the degree of substitution of the carboxymethyl groups. If the degree of substitution is 0.2, its molecular mass is 178.14 g/mol. If this degree reaches 1.5, the mass of the glucose building block increases to 282.18 g/mol. Food-grade macromolecules have a molecular mass of at least 17,000 g/mol, which corresponds to approximately 100 degrees of polymerization (DP). Additionally, according to the provided relationship between the degree of modification and its sodium content, the amount of sodium for each DS is ten times higher [4–7, 14].

Table 8.1 Relationship between viscosity, molecular weight, and DP (degree of polymerization) [14]

CMC—viscosity (mPa s) (2%, Brookfield)	Molecular weight (g mol^{-1})	Polymerization degree (DS = 0.7)
Low (< 100)	90,000	400
Medium (< 5000)	250,000	1100
High (> 20,000)	700,000	3200

Table 8.2 Relationship between degree of substitution and sodium content of a CMC [4, 14]

Substitution degree (DS)	0.1	0.2	0.3	0.4	0.5	0.6	0.7	0.8
Sodium content (wt%)	1.35	2.58	3.69	4.72	5.66	6.53	7.33	8.08
Substitution degree (DS)	0.9	1.0	1.1	1.2	1.3	1.4	1.5	
Sodium content (wt%)	8.77	9.42	10.02	10.59	11.12	11.62	12.09	

In the food industry, Na-CMC is used to enhance the hydrogel properties of products and manage the separation of ingredients [8]. In the cosmetics industry, Na-CMC is added as a thickening agent in the composition of toothpastes and creams [3, 9]. In the pharmaceutical field, Na-CMC is used in the production of hydrogels and as a component of drug delivery matrices in bioengineering (for controlled release of active substances in medications) [10–13].

Household products contain many ingredients (such as silicates, oils, surfactants, proteins, polyols, sugars, and salts) that interact with Na-CMC, potentially forming new compounds. These compounds have been found to significantly differ from the properties of simple single-component solutions. Therefore, studying the impact of Na-CMC solutions' rheological descriptions is of great practical importance [10–14, 17].

Like other polysaccharide solutions, Na-CMC solutions can be used as drug delivery systems due to their high viscosity, which can be achieved by selecting the right components and concentrations [16, 17]. The results of this study can also be practically applied in shaping products in the food and cosmetics industries [16, 17].

A deeper analysis of Na-CMC's industrial applications, understanding its properties, and monitoring market trends can stimulate scientific and technological research related to this polymer. Moreover, its application in new fields is expected to expand in the future, making it an increasingly valuable and in-demand resource [16, 17].

The properties of Na-CMC are intricately linked to its molecular structure, with factors such as the amount of carboxymethyl groups and the length of the molecular chain defining its physical and chemical properties. The length of its molecular chain and the distribution of carboxymethyl groups ensure the product's water solubility, stability, and strength, thereby expanding its range of applications [13, 16].

Currently, the level of study on the physical properties of locally produced Na-CMC is insufficient. Additional scientific and technological research related to this polymer could enhance understanding and expand its application possibilities in new fields. Future applications of this material in fields such as biotechnology and environmental protection are anticipated, which would transform it into an even more valuable and sought-after resource. Thus, a deeper exploration of Na-CMC's properties and its potential in new scientific and practical applications is necessary [10, 12].

8.2 Materials

Sodium carboxymethylcellulose (Na-KMS)—SSt 5.588-79 and BA 6-05-386-80. Purified sodium carboxymethyl cellulose, a product of the Namangan Chemical Plant, obtained by the method of heterogeneous solid-phase esterification of sulfite wood cellulose with monochloroacetic acid (MCA) of the following structure was used as the main object of study (Fig. 8.3) [11].

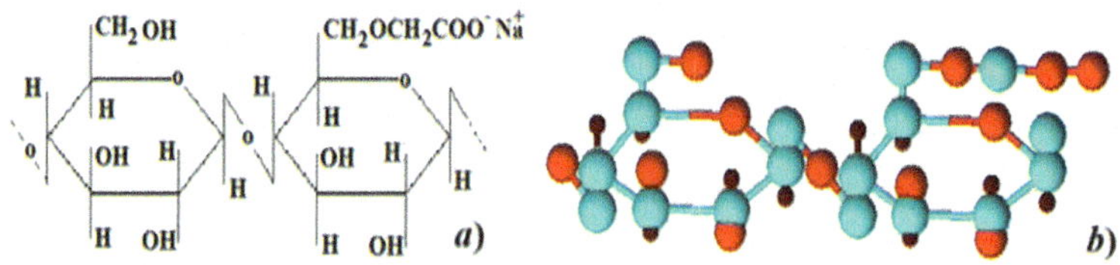

Fig. 8.3 Chemical formula (**a**) and model structure of Na-KMS (**b**) [11]

Using Ubbelohde capillary viscometry and based on the Mark-Houwink equation, we assessed the relative molecular mass of Na-CMC to be around 0.1×10^6, categorizing it as a relatively light polymer [16].

Sodium carboxymethyl cellulose is a local product manufactured at the Namangan chemical plant. Prior to sample preparation, it is purified in a Soxhlet apparatus to remove any potential foreign substances that could have been incorporated due to various interactions [11].

Sample Preparation. A homogeneous 8% concentration sample of sodium carboxymethyl cellulose solution was prepared at room temperature (25 °C) and under specially arranged conditions, and its rheological properties were studied [13].

8.3 Methods

Rheometry is an experimental method used to determine the rheological properties of materials by studying the interrelationship between force, deformation, and time through changes in flow and deformation. It is essential to understand the flow and viscosity properties of solutions and suspensions to grasp their rheological characteristics. Research Equipment. With the advancement of science and technology, the sensitivity and inspection capabilities of devices are continuously improving. The multifunctional Anton Paar MCR 92 (Modular Compact Rheometer) device allows for more accurate and comprehensive testing of the rheological properties of Na-CMC solutions compared to previous methods. The MCR 92 is one of the modern devices capable of operating continuously (24 h a day) without stopping. The device can generate rotational and oscillatory torques ranging from 1 to 125 mN m, with a speed range (CSS/CSR) from 0.001 to 1500 rpm, and is designed to operate under various experimental conditions within a temperature range of − 50 to + 400 °C [14–17].

Under external forces, a transverse stress σ is generated, defined as the force F acting per unit area S [16].

$$\sigma = \frac{F}{S} \tag{8.1}$$

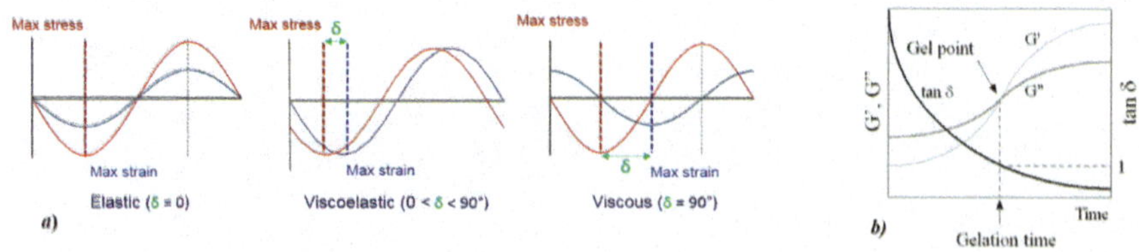

Fig. 8.4 **a** A graph showing the phase delay between loading and deformation reactions, **b** a graph depicting the point of gel formation [17]

During flow, liquids form interlayer movements. Friction between these layers in liquid flow leads to a decrease in both the velocity of the fluid and its kinetic energy. This loss is primarily associated with the fluid's internal or dynamic viscosity η [4, 16]. The relationship between transverse stress and the rate of shear deformation is linked to the fluid's viscosity [14, 16].

$$\eta = \frac{\sigma}{\dot{y}} \tag{8.1}$$

When a periodic external load is applied to a solution, the physical parameters of the sample undergo dynamic changes. These changes in physical properties are determined by the frequency of the applied external load and the specific relaxation time of the system. When the sample is subjected to periodic loading, a phase difference between stress and deformation occurs. In polymer solutions, the amount of this phase delay is quantitatively determined by the loss tangent, which is the ratio of the loss modulus G″ to the storage modulus G′ (Fig. 8.4) [13, 14]:

$$\tan \delta = \frac{G''}{G'} \tag{8.1}$$

The critical point where the values of the storage and loss moduli are equal (tan δ = 1) is called the gel point. In the gel phase, even though the mixture remains in a liquid state, the molecules are significantly more strongly bonded than in other phases. By analyzing the phase difference between deformation and stress in a polymer solution, it is possible to determine the molecular conformations within the solution and the material's response to external forces. This analysis helps to understand how the molecules in the solution respond to external forces and assess the stability of the material's deformation under such conditions. As the temperature changes, the molecular conformations alter, and the rheological process varies accordingly. Studying the impact of temperature on rheological properties helps to deepen understanding of how the material reacts to various external influences.

FTIR. The structure of the products was determined using IR spectroscopy and electron microscopy. Infrared spectra were recorded in the range of 400–4000 cm^{-1} on "NICOLET Magna-560 IR" and "Specord-75IR" spectrophotometers (Carl Zeiss, GDR). For IR spectroscopy, samples were prepared as KBr pressed pellets, films on

KRS-5 plates, and films of 8–12 μm thickness. Films on KRS-5 plates were obtained by evaporating the solvent at room temperature (295–297 K) [11].

XRD. The device used for measuring X-ray diffraction is an X-ray diffractometer, which operates with X-ray radiation originating from a copper (Cu) source. This device works at 40 kV voltage and 30 mA current, offering the capability to rotate samples within an angular range from 4° to 80°. The sample rotation speed is 2°/min, allowing for accurate and repeatable measurements. The diffraction of X-rays through crystalline structures is carried out according to the laws of X-ray diffraction [11].

8.4 Results and Discussion

The following image displays the characteristic absorption lines in the IR spectrum of a sodium carboxymethyl cellulose (Na-CMC) sample. Na-CMC is a polymeric material whose molecular structure is characteristic of water-soluble polyelectrolytes. Below is the IR spectrum provided (Fig. 8.5).

Analyzing the main bonds in the Na-CMC molecule provides critical information about the material's structure:

1. **OH group (3400–3200 cm^{-1})**. A region with high absorption associated with hydroxyl (OH) groups, indicating a significant presence of these groups in the polymer.
2. **CH_2 va CH_3 (3000–2800 cm^{-1})**. These bonds are characteristic of methylene (CH_2) and methyl (CH_3) groups, primarily reflecting the hydrocarbon chains in Na-CMC.
3. **C=O carboxyl group (1600 cm^{-1})**. This distinct area arises from the C=O bonds of carboxyl groups. It is a primary functional group of the Na-CMC molecule, indicating the presence of carboxymethyl groups.

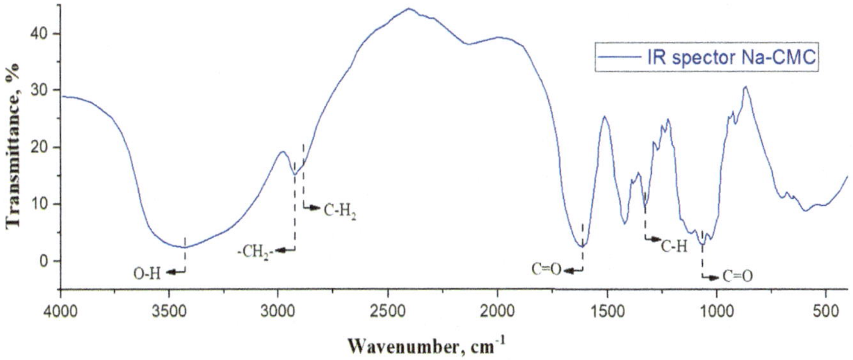

Fig. 8.5 Na-CMC IR spectra

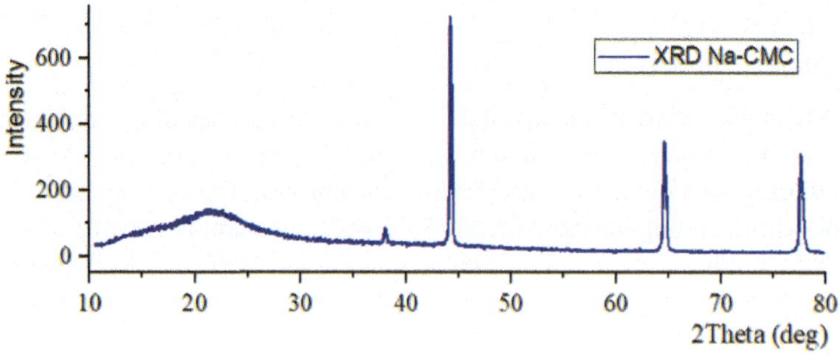

Fig. 8.6 Na-CMC's X-ray diffractometer

4. **C-O stretching (1000–1300 cm^{-1})**. Includes the C-O stretches of carboxymethyl groups. These are crucial chemical groups that contribute to the molecule's water solubility. These spectral features are essential for a deeper understanding of Na-CMC's chemical composition and molecular structure.

These spectral features are essential for a deeper understanding of Na-CMC's chemical composition and molecular structure.

The XRD (X-ray Diffraction) spectrum of the Na-CMC sample is shown. XRD spectra are used to determine the crystalline or amorphous structure of the sample and to study the material's phase composition. The following description is based on the analysis of the XRD spectrum for Na-CMC (Fig. 8.6).

This graph indicates a well-ordered crystalline lattice and a high degree of crystallinity.

The following graph shows the rheological properties obtained for the Na-CMC solution, specifically the relationship between viscosity and shear rate. The data displayed illustrates the relationship between the viscosity of the polymer solution and the edge speed, which is crucial for understanding Na-CMC's temperature-dependent properties. These properties are key in determining the material's suitability for various industrial applications. The vertical axis represents viscosity (Pa s), and the horizontal axis represents angular velocity (s^{-1}). Lines for different temperatures are shown, through which the changes in Na-CMC with temperature can be observed. Each line is measured at a specific temperature, and a decrease in viscosity with increasing temperature is noted. This phenomenon can be explained by the decrease in intermolecular bonding strength at higher temperatures, thereby facilitating easier movement of the solution (Fig. 8.7).

On the left side, high viscosity is observed at low angular velocities. As the angular velocity increases, the viscosity sharply decreases, a phenomenon known as "shear thinning." This phenomenon indicates that the internal resistance of the material decreases as it is forced to move faster, facilitating easier fluid movement.

The nearly horizontal line in the lower part of the graph indicates that the viscosity remains almost constant with increasing angular velocity, demonstrating Na-CMC's

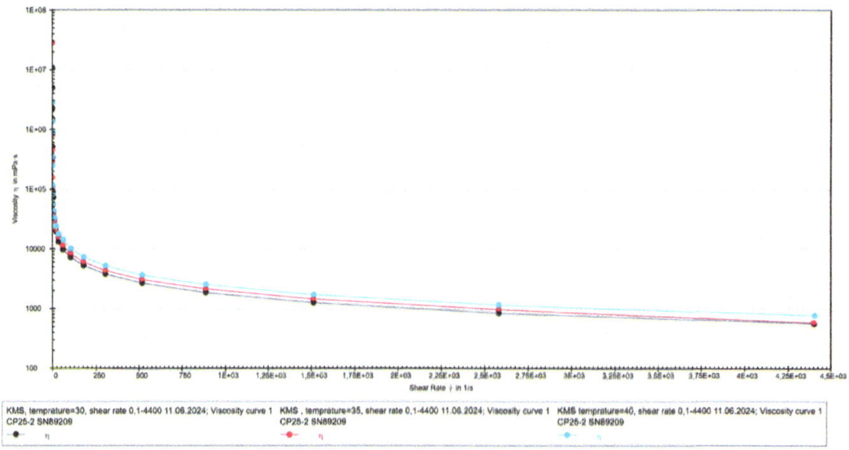

Fig. 8.7 Dependency of the internal friction coefficient on angular velocity at different temperatures

high stabilization properties. This condition signifies the structural stability of the material and its ability to perform over a long duration in applied environments.

The following graph describes how the shear stress for Na-CMC changes at various temperatures. The horizontal axis represents angular velocity (s^{-1}), while the vertical axis shows shear stress (Pa). Three different colored lines represent data at various temperature levels. The blue line shows data at the lowest temperature, the navy line at a medium temperature, and the red line at the highest temperature.

Each line's appearance demonstrates the following trend: shear stress increases with increasing shear rate. However, the rate of increase in shear stress decreases with increasing temperature. This phenomenon may be related to the decrease in intermolecular bonding strength and easier molecular movement at higher temperatures (Fig. 8.8).

Analyzing the rheological properties of the material at a temperature of 25 °C, we attempt to provide a deeper understanding of its industrial applications and mechanical stability (Table 8.3).

Analyzing the results graphically aids in comprehending the overall process. The following graph displays the dependence of the Loss Factor on Angular Frequency in the rheological analysis of the Na-CMC solution. The loss factor indicates the material's response level to external stress, specifically how much energy is dissipated in the form of heat (Fig. 8.9).

The graph shows the variations of the loss factor with angular frequency. As angular frequency increases, the loss factor initially rises quickly, and then experiences several fluctuations at certain points, and finally, it sharply decreases at higher frequencies.

Initial Rise (0–100 rad/s): In this region, the loss factor increases rapidly, indicating the material's good ability to absorb energy at low frequencies. This phenomenon may be related to the flexibility of the material's internal structures and energy dissipation.

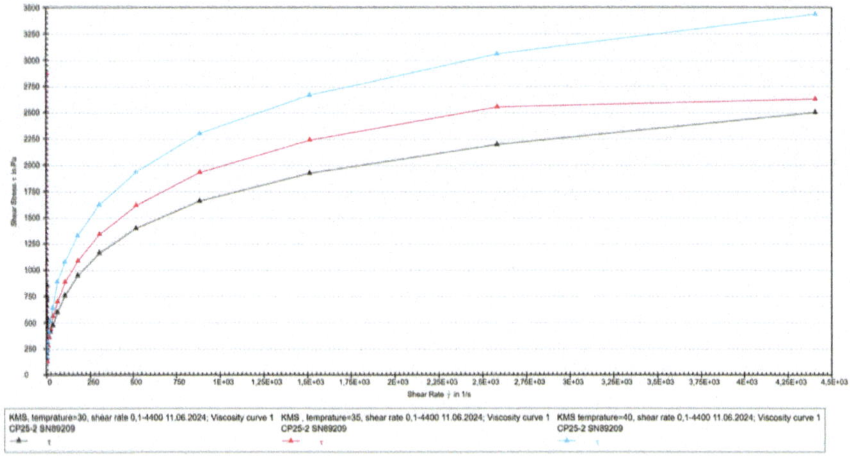

Fig. 8.8 Dependency of the shear rate on temperature

Table 8.3 Rheological analysis results of Na-CMC at 25 °C

No.	Angular frequency (rad/s)	Storage modulus (Pa)	Loss modulus (Pa)	Loss factor	Shear strain (%)	Shear stress (Pa)	Torque (mN m)
1	0.314	1272.1	623.42	0.49	0.101	1.4333	0.005848
2	0.521	1.60E+05	1.32E+05	0.828	0.0146	30.402	0.12403
3	0.866	1.30E+06	3.88E+05	0.297	0.977	13,297	54.25
4	1.44	29,225	15,515	0.531	0.581	192.35	0.78475
5	2.38	23,167	9834.7	0.425	1.01	253.49	1.0342
6	3.96	1943.3	2807	1.444	2.88	98.46	0.4017
7	6.57	11,024	8197.4	0.744	0.833	114.39	0.4667
8	10.9	6922.3	6528.8	0.943	0.972	92.521	0.37747
9	18.1	6073.1	7032.4	1.158	0.985	91.549	0.37351
10	30	9409.6	9061.9	0.963	0.862	112.61	0.45943
11	49.9	15,754	10,892	0.691	0.963	184.41	0.75235
12	82.8	8700.8	9724.7	1.118	0.99	129.15	0.5269
13	137	9354.9	9832.6	1.051	0.997	135.31	0.55205
14	228	8571.3	10,161	1.185	1	132.96	0.54246
15	379	3968.2	10,204	2.571	0.999	109.39	0.44631
16	628	12,917	0.64585	0	1	129.12	0.52679

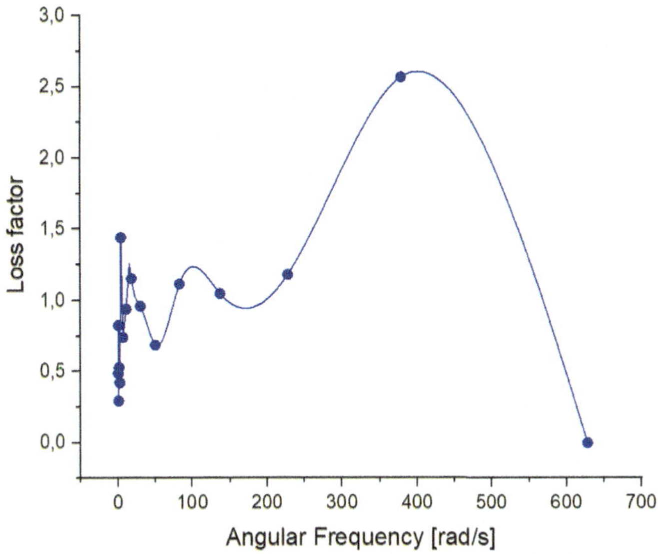

Fig. 8.9 Variation of the loss factor at different angular velocities

Fluctuations (100–300 rad/s): During this period, the graph shows that the loss factor undergoes several fluctuations. This may indicate that the material enters a resonance state at certain frequencies, which stems from the interactions within the material's internal structures at these specific frequencies.

Maximum Value (300–400 rad/s): The loss factor reaches its highest value, indicating the most significant energy dissipation within this frequency range. This suggests that the material loses the most energy and concurrently shows a decrease in operational efficiency at these frequencies.

Sharp Decline (400–600 rad/s): The loss factor decreases sharply, indicating a significant reduction in the material's ability to absorb energy at high frequencies. This reflects the inability of the molecular bonds within the material to efficiently dissipate energy at high speeds.

8.5 Conclusions

The XRD analysis of the Na-CMC (sodium carboxymethyl cellulose) sample determined its degree of crystallinity. The most significant peaks in the sample are located at 21°, 37°, 44°, 64°, and 78° 2Theta, and their overall intensity indicates the level of crystallinity in the sample, which appears to be high.

The rheological properties of the Na-CMC solution were thoroughly studied, allowing for the identification of critical properties of this polymer solution at various temperatures. Based on data obtained from samples prepared at a concentration of 8% and a temperature of 25 °C, the following important conclusions were made.

Viscosity and angular velocity: The Na-CMC solution exhibits high viscosity, which sharply decreases as the shear rate increases (shear thinning characteristic). This property enhances the efficiency of processing the polymer solution.

Gel point: The gel point was observed at a frequency of 30 rad/s with storage modulus G′ at 9409.6 Pa and loss modulus G″ at 9061.9 Pa, where the tan(δ) value between G′ and G″ is close to 1. These changes indicate structural transformations in the Na-CMC solution. Effect of temperature: An increase in temperature was observed to decrease the viscosity and loss modulus, indicating the temperature sensitivity of the Na-CMC molecular structure and its improved performance in warmer environments.

References

1. C. Xiao, Y. Gao, Preparation and properties of physically crosslinked sodium carboxymethylcellulose/poly(vinyl alcohol) complex hydrogels. J. Appl. Polym. Sci. **107**, 1568–1572 (2007)
2. A. Benslimane, I.M. Bahlouli, K. Bekkour, D. Hammiche, Thermal gelation properties of carboxymethyl cellulose and bentonite-carboxymethyl cellulose dispersions: rheological considerations. Appl. Clay Sci. **132–133**, 702–710 (2016)
3. A. Kukrety, R.K. Singh, P. Singh, S.S. Ray, Comprehension on the synthesis of carboxymethylcellulose (CMC) utilizing various cellulose rich waste biomass resources. Waste Biomass Valoriz. **9**, 1587–1595 (2018)
4. M.S. Yeasmin, M.D.I.H. Mondal, Synthesis of highly substituted carboxymethyl cellulose depending on cellulose particle size. Int. J. Biol. Macromol. **80**, 725–731 (2015)
5. J.S. Behra, J. Mattsson, O.J. Cayre, E.S.J. Robles, H. Tang, T.N. Hunter, Characterization of sodium carboxymethyl cellulose aqueous solutions to support complex product formulation: a rheology and light scattering study. ACS Appl. Polym. Mater. **1**, 344–358 (2019)
6. A. Benchabane, K. Bekkour, Rheological properties of carboxymethyl cellulose (CMC) solutions. Colloid Polym. Sci. **286**, 1173–1180 (2008)
7. M. Helmut, Rheological measurements and structural analysis of polymeric materials. Polymers **13**, 1123 (2021)
8. L. Sangroniz, M. Fernandez, A. Santamaria, Polymers and rheology: a tale of give and take. Polymer **271**, 125811 (2023)
9. S.Y. Inagamov, U.A. Asrorov, N.T. Qodirova, O.T. Yomgirov, G.I. Mukhamedov, Development of interpolymer complexes based on sodium carboxymethylcellulose and polyacrylamide. II International Scientific and Practical Conference Technologies, Materials Science and Engineering, AIP Conf. Proc. **2999**, 020049 (2023)
10. P. Wagner, S. Rózanska, E. Warmbier, A. Frankiewicz, J. Rózanski, Rheological properties of sodium carboxymethylcellulose solutions in dihydroxy alcohol/water mixtures. Materials **16**, 418 (2023)
11. S.Y. Inagamov, U.A. Asrorov, E.B. Xujanov, Structure and physico-mechanical properties of polyelectrolyte complexes based on sodium carboxymethylcellulose polysaccharide and polyacrylamide. East Eur. J. Phys. **4**, 258–266 (2023)
12. M.L. Willams, R.F. Landel, D.J. Ferry, *The temperature Dependence of Relaxation Mechanisms in Amorphous Polymers and Other Glass-forming Liquids*. Contribution from the Department of Chemistry, University of Wisconsin (1955)

13. T. Wüstenberg, Chapter 10. Sodium carboxymethylcellulose, *Cellulose and Cellulose Derivatives in the Food Industry* (2014)
14. M. Puhringer, C.H. Paulik, K. Bretterbauer, Synthesis and characterization of poly-acrylamide-based biomimetic underwater adhesives. Monatshefte für Chemie—Chem. Month. **154**, 503–513 (2023)
15. Z. Su, Ch. Li, G. Jiang, J. Li, H. Lu, F. Guo, *Rheology and Thixotropy of Cement Pastes Containing Polyacrylamide* (Hindawi, Geofluids, 2022)
16. H. Murata, Rheology—theory and application to biomaterials. Polymerization (2012)
17. H. Ramli, N.F.A. Zainal, M. Hess, Ch.H. Chan, Basic principle and good practices of rheology for polymers for teachers and beginners. Chem. Teach. Int. **4**(4), 307–326 (2022)

Chapter 9
A New Approach to the Analysis of Nuclear Reactions Under the Influence of High-Energy Particles on a Stationary Target Nucleus in the Physics of Resonant Nuclear Reactions

Ibratjon Aliyev, G'olibjon Qo'ldashov, Sultonali Abdurakhmonov, Inomjon Bilolov, and Sharobiddin Isroilov

Abstract The paper presents a new form of complex analysis of nuclear reactions brought to a state of resonance. This state of nuclear reactions is investigated according to the model of physics of resonant nuclear reactions. The study examines each channel of the nuclear reaction, calculates the nuclear effective cross-section of each of the channels before and after reduction. During the analysis, attention is drawn to the theoretical study of the energy characteristics of the products of each of the reaction channels, with a percentage ratio. The temperature gradients of the nuclei and targets formed at different stages of the reaction are determined. As a result, a complete model has been formed to create a complete description of the nuclear reaction being produced and analyzed, which makes it possible to theoretically predict the outcome of any nuclear reaction. The analysis concludes with the final result in the form of all parameters of the nuclear reaction products at all stages.

I. Aliyev (✉)
SRI "PRNR", Electron Laboratory LLC, Margilan, Fergana Region, Republic of Uzbekistan
e-mail: alievibratzon12@gmail.com

G. Qo'ldashov
National Research of "Renewable Energy Sources" Under the Ministry of Energy of the Republic of Uzbekistan, Tashkent Region, Tashkent, Republic of Uzbekistan

S. Abdurakhmonov · I. Bilolov · S. Isroilov
Ferghana State Technical University, Fergana, Fergana Region, Republic of Uzbekistan

© The Author(s), under exclusive license to Springer Nature Switzerland AG 2025
A. Gibadullin et al. (eds.), *Proceedings of the 4th International Scientific and Practical Symposium on Materials Science and Physics*, Springer Proceedings in Physics 318,
https://doi.org/10.1007/978-3-031-94805-3_9

9.1 Introduction

The implementation of a nuclear reaction is the bombardment of a beam of charged particles—ionized nuclei, leptons, elementary particles of other categories with selected parameters of the target nuclei. To date, in order to study nuclear reactions occurring during bombardment by high-energy charged particles of stationary target nuclei, experimental research is primarily being conducted using charged particle accelerators at various beam energies with different levels of monochromatization [1]. As a result, at stages with increased monochromatization, the transition of the entire reaction to a state of resonance is observed [2, 3], while the accelerator can act in the form of various structures, depending on the specified parameters.

A common case when using this method is the use of several channels of acceleration lines, where the beam is separated by a separator, often electromagnetic, into separate beams that collide with the target in different chambers [4, 5]. Thus, to calculate the nuclear effective cross section of the reaction, it is necessary to use a special chamber with a dense insulating, more often lead shell and separate detectors. Cylindrical chamber designs are used to analyze the reaction results with the additional possibility of studying the Rutherford scattering effect [4, 6–8].

Based on this, it is clearly seen that these reactions require a sufficient amount of time to prepare and direct the corresponding energy costs. Therefore, sufficient preparatory time, material and energy costs are required to study the interaction of beams with target nuclei in various combinations [9, 10]. After the experiment, the results are processed using the available theoretical analysis methodology, which is described in many research papers [11, 12]. However, in order to accelerate research in the field of nuclear reactions, a new approach is needed that outputs results when certain initial and boundary conditions change—the parameters of the accelerator, beam and target, therefore, the creation of a new approach that accelerates the obtaining of results in nuclear reactions is relevant.

A separate problem today is the complexity of studying a new type of reaction, resonant nuclear reactions. Such reactions can be carried out on accelerators with high beam monochromacy [3, 13]. So today, on cyclotrons, synchrotrons, phasotrons, and linear accelerators, monochromaticity reaches 5 keV at a beam energy of 20 MeV [2, 3, 5, 14]. An increase in this degree is possible with rigid separation, which leads to a large loss of beam current by 70–80% [16–18, 21]. Van de Graaf accelerators with a beam monochromaticity of 1 keV at the same maximum beam energy are considered to be the most accurate [19, 20].

An increase in this degree is possible with a more accurate selection of energy using nano-electrodes and special monochromotic devices that are at the design stage [2, 3, 22–25]. However, it is important to create a mathematical model beforehand that allows not only predicting the implementation of nuclear reactions, but also all their varieties, including resonant nuclear reactions with great accuracy. Based on this, the task is also relevant.

9.2 Materials and Methods of Research

In the course of the research, the method of solving boundary and initial conditions in the form of accelerator system parameters, the method of classification, analysis, and mathematical modeling was used.

9.3 Research

The analysis of nuclear reactions can be carried out in a variety of ways and using different algorithms; however, the purpose of the analysis remains unchanged—to create a complete picture of the analyzed nuclear reaction. In the early analysis model [1], initially one nuclear reaction was isolated, one specific channel of it was analyzed—its energy characteristic, nuclear effective cross-section, percentages and other parameters were determined, and only after the study moved to the stage of analyzing other channels of the nuclear reaction, and at the same time the initially analyzed channel remained isolated. Each time, the chosen channel in the exo-energy reactions had a large nuclear effective cross-section, which seemed inexplicable.

The problem was that the initial channel was selected more often for the energy of the Coulomb barrier, as a result, it was observed that the reaction was carried out in an area with a large percentage, which turned out to be the main channel with a large reaction yield, since the main criterion for selected nuclear reactions in early analyses relative to the previous algorithm was the efficiency in the reaction output. In the new model of reaction analysis, the solution to this problem plays a special role, presented in the following form.

Suppose that a nuclear reaction c with its own channels (9.1) is initially set.

$$a_{x_1}^{y_1} + A_{x_2}^{y_2} \rightarrow \begin{cases} B_{1(x_3)}^{1(y_3)} + b_{1(x_4)}^{1(y_4)} \\ B_{2(x_3)}^{2(y_3)} + b_{2(x_4)}^{2(y_4)} \\ \ldots \\ B_{n(x_3)}^{n(y_3)} + b_{n(x_4)}^{n(y_4)} \end{cases} \tag{9.1}$$

Which can also be converted to (9.2).

$$a_{x_1}^{y_1} + A_{x_2}^{y_2} \rightarrow \begin{cases} B_{x_1+x_2-1(x_4)}^{y_1+y_2-1(y_4)} + b_{1(x_4)}^{1(y_4)} \\ B_{x_1+x_2-2(x_4)}^{y_1+y_2-2(y_4)} + b_{2(x_4)}^{2(y_4)} \\ \ldots \\ B_{x_1+x_2-n(x_4)}^{y_1+y_2-n(y_4)} + b_{n(x_4)}^{n(y_4)} \\ \ldots \\ B_{x_1+x_2}^{y_1+y_2} \end{cases} \tag{9.2}$$

Initially, the analysis assumes the determination of the radius of the target core being bombarded (9.3), in order to further determine the Coulomb incoming barrier (9.4).

$$R_{A_{x_2}^{y_2}} = 1.4 * A^{\frac{1}{3}} = 1.4 * (y_2)^{\frac{1}{3}} \tag{9.3}$$

$$B_k = \frac{\left|q_{a_{x_1}^{y_1}}\right|\left|q_{A_{x_2}^{y_2}}\right|}{R_{A_{x_2}^{y_2}}}\frac{e^2}{\hbar c}\hbar c \tag{9.4}$$

After determining the Coulomb incoming barrier, which can most often be higher than the threshold of a nuclear reaction, the energy of the incoming particle is determined, and as the energy approaches, it becomes possible to increase the probability of a larger number of nuclear reactions. However, in each case, more than one nuclear reaction channel is considered (9.1), among which there are reaction channels that are maximally probabilistic (9.5).

$$a_{x_1}^{y_1} + A_{x_2}^{y_2} \rightarrow \begin{cases} B_{x_1+x_2-1}^{y_1+y_2-1} + p_1^1 \\ B_{x_1+x_2+1}^{y_1+y_2} + e_{-1}^0 \\ B_{x_1+x_2-1}^{y_1+y_2} + e_{+1}^0 \\ B_{x_1+x_2-1}^{y_1+y_2-2} + H_1^2 \\ B_{x_1+x_2-1}^{y_1+y_2-3} + H_1^3 \\ B_{x_1+x_2-2}^{y_1+y_2-4} + He_2^4 \\ \quad \ldots \\ B_{x_1+x_2}^{y_1+y_2} \end{cases} \tag{9.5}$$

Such channels include channels with outgoing elementary particles or relatively light nuclei, among which one can single out a proton as the nucleus of a hydrogen atom, a deuteron and a triton as the nuclei of deuterium and tritium, respectively, as well as alpha particles.

But also, together with the above-mentioned probabilistic reaction channels, it is possible to carry out reactions with the departure of lighter reactions if, compared with heavy nuclei, the departing nucleus is with sufficiently small masses, however, it is the channels with the departing particles in (9.5) that are most often determined. In the early analysis, the situation with energies not reaching the surface was not considered. The value up to the Coulomb barrier, however, in this case, if necessary, this moment is considered.

To study this variation, a Rutherford scattering variant is demonstrated, where the scattering angle (9.6) is initially determined, which determines the ratio from the cross-sectional area of the incident beam to the total area of a sphere with a radius equal to the beam radius, which makes it possible to determine in steradians the angle from which the beam is directed towards the core, demonstrating a one-sided The direction of the beam is different than in the cases of special accelerators with multiple accelerator systems, which include a system for starting a thermonuclear reaction with a large angle.

$$\theta = \frac{S_{a_{x_1}^{y_1}}}{S_O} \tag{9.6}$$

Also, at the same time, the boundary velocity of the incoming beam (9.7) is determined, which is calculated on the condition that the energy of the beam is exactly equal to the energy of the Coulomb barrier, as a result of which, after passing, the particle simply would not have kinetic energy left.

$$v'_{a_{x_1}^{y_1}} = c\sqrt{1 - \left(\frac{1}{1 + \frac{B_k}{m_{a_{x_1}^{y_1}} c^2}}\right)^2} \tag{9.7}$$

In addition to the factor of determining the boundary velocity, it is worth adding that the beam velocity cannot be exactly equal to the boundary velocity due to the impossibility of ensuring zero monochromatization in the accelerator, as well as due to the law of the impossibility of having zero kinetic energy in particles. After determining the necessary ratios in (9.6) and (9.7), they are substituted into the expression for calculating the effective cross section, which is derived in Eq. (9.8).

$$\frac{d\sigma}{d\Omega} = \left(\frac{Z_1 Z_2 e^2}{2mv^2_{a_{x_1}^{y_1}}}\right)^2 \frac{1}{\sin^2\frac{\theta}{2}} => \sigma = \left(\frac{Z_1 Z_2 e^2}{2mv^2_{a_{x_1}^{y_1}}}\right)^2\Bigg|_0^{v_{a_{x_1}^{y_1}}} \left(-\frac{2\cot^3\frac{\theta}{2}}{3} - 2\cot\frac{\theta}{2}\right)\Bigg|_0^{\theta'} \tag{9.8}$$

Along with this, in order to determine the minimum distance between the incoming beam and the core (9.10), an additional parameter of the electromagnetic diameter of the core (9.9) is calculated.

$$b = \frac{Z_{A_{x_1}^{y_1}} e^2}{T_{a_{x_1}^{y_1}} \tan\frac{\theta}{2}} \tag{9.9}$$

$$r_{min} = \frac{Z_{A_{x_1}^{y_1}} e^2}{T_{a_{x_1}^{y_1}}} + \frac{b}{\cos\frac{\theta}{2}} = \frac{Z_{A_{x_1}^{y_1}} e^2}{T_{a_{x_1}^{y_1}}}\left(1 + \frac{1}{\sin\frac{\theta}{2}}\right) \tag{9.10}$$

Thus, as a result of the initial part of the analysis, it was possible to determine the effective cross-section, which is measured in bars (9.8) and represents the probability of a barrier with the calculation of the distance at which the incoming particle approaches the nucleus without interacting with it (9.10). The data determined in this way is quite exhaustive to describe a situation where elastic interaction between particles is carried out without entering into a reaction. The only exceptions are thermonuclear reactions, in which, due to universal ionization and a number of other factors, the presence of the Coulomb barrier is ignored, acting according to separate private algorithms.

In this analysis, the situation with the incoming particle beams in the accelerator is considered, as a result of which the probabilities can be determined according to the previously presented patterns. After all possible channels of nuclear reactions have been identified, but most often more probabilistic in the first degree with the release of light particles, the output of such channels (9.11) and the thresholds of such reactions (9.12) are calculated.

$$Q_i = \left(M_{a_{x_1}^{y_1}} + M_{A_{x_2}^{y_2}} - M_{B_{i(x_2)}^{i(y_2)}} - M_{b_{i(x_2)}^{i(y_2)}} \right) c^2 \tag{9.11}$$

$$T_i = |Q_i| \left(1 + \frac{M_{a_{x_1}^{y_1}}}{M_{A_{x_2}^{y_2}}} + \frac{|Q_i|}{2M_{A_{x_2}^{y_2}} c^2} \right) \tag{9.12}$$

After the reaction channel is calculated in each of the cases, the second-degree probabilistic channels of nuclear reactions can be determined. Such channels have a positive output of nuclear reactions, that is, they are exo-energetic, if any, or if all reactions are endo-energetic, then, provided that the required energy level is reached by the incoming beam to satisfy such a reaction, that is, if there is a possibility of releasing the energy that the reaction takes, the probability of the reaction it is distributed between them.

Determining the output of the reaction channels allows not only to characterize each of the channels, but also to determine the probability of what percentage of the total set of particles will be directed to implement one or another probabilistic nuclear reaction channel. This probability can be determined by (9.13) as a percentage, but in the future, in each of the cases, the nuclear effective cross-section of each of the nuclear reaction channels must be determined, for which the density of the target nuclei (9.14) is initially calculated, which is the number of target nuclei per cubic meter, which can be It is defined as the ratio of the target density to the mass of the nuclei in kilograms.

$$\eta_i = \frac{Q_i}{\sum_{k=1}^{n} Q_k} \tag{9.13}$$

$$n_{A_{x_2}^{y_2}} = \frac{\rho_{A_{x_2}^{y_2}}}{M_{A_{x_2}^{y_2}}} \tag{9.14}$$

In the classical form of the analysis, the derivation of the expression was partially presented, the relationship between the initial number of incoming particles in the beam to the number of particles involved in the reaction and in which the target thickness and nuclear density (9.14) are involved, together with the value of the nuclear effective cross-section, which can also be derived in (9.15).

$$N = N_0 e^{-n_{A_{x2}^{y2}} \sigma_i dx_{A_{x2}^{y2}}} => \sigma_i = -\frac{\ln n_i}{n_{A_{x2}^{y2}} dx_{A_{x2}^{y2}}} \tag{9.15}$$

Thus, for each of the channels, a certain set is calculated in the form of a sequence of values of the nuclear effective cross section (9.16).

$$\sigma_1 < \sigma_2 < \ldots < \sigma_{n-1} < \sigma_n \tag{9.16}$$

Each of these values is determined directly for the case when the energy of the impinging particle is directly selected for one or another channel to which the value of the nuclear effective cross-section belongs in the sequential set (9.16). It is worth noting that each of the values of the nuclear effective cross-section divides the line of values in bars into certain areas. So, in order to determine exactly the situation with the nuclear effective cross-section and the percentage distribution in the case of strictly defined energy values in this case, it is necessary to act according to a different algorithm, which in this system will restore order in the case when the nuclear effective cross-section in the case under consideration falls into one of the areas, which makes this area into which The value was more probabilistic in the immediate case under consideration.

Thus, in order to achieve the optimal case, the energy of the incident particle is chosen as the sum of the Coulomb incoming barrier and a certain value, which makes the beam energy equal to the value that the accelerator can achieve with its own monochromatic value (9.17).

$$E_{a_{x1}^{y1}} = B_k + E_u \tag{9.17}$$

Then, after determining the energy (9.17), the velocity of the particle (9.18) is calculated, which cannot be confused with the boundary values of velocity, as can be seen from the formula, after which this value is substituted into the expression of momentum (9.19) according to the theory of relativity with known mass values, due to sufficiently large velocity values, and after which this value is inserted into the expression definitions of the de Broglie wave (9.20) for an incoming particle.

$$v_{a_{x1}^{y1}} = c\sqrt{1 - \left(\frac{1}{1 + \frac{E_{a_{x1}^{y1}}}{m_{a_{x1}^{y1}} c^2}}\right)^2} \tag{9.18}$$

$$p_{a_{x1}^{y1}} = \frac{m_{a_{x1}^{y1}} v_{a_{x1}^{y1}}}{\sqrt{1 - \frac{v^2_{a_{x1}^{y1}}}{c^2}}} \tag{9.19}$$

$$\lambda_{a_{x1}^{y1}} = \frac{h}{p_{a_{x1}^{y1}}} \tag{9.20}$$

After the necessary components are determined, the nuclear effective cross-section (9.21) can be calculated, which is an area larger than the cross-section of the nucleus and in which the incoming particle also interacts without even touching the nucleus.

$$\sigma = \pi\left(R_{A_{x_2}^{y_2}} + \lambda_{a_{x_1}^{y_1}}\right)^2 \tag{9.21}$$

Then, after that, the percentage (9.22) belonging to the maximum percentage value can be determined, taking into account the factor that the numerator refers to the number of particles that did not interact.

$$\eta = \frac{N}{N_0} = e^{-n_{A_{x_2}^{y_2}}\sigma_i dx_{A_{x_2}^{y_2}}} \tag{9.22}$$

And now, to determine which channel of a nuclear reaction such a reaction belongs to, the effective cross-section calculated in (9.21) is substituted into a certain area, into which the area of nuclear effective cross-sections was previously divided, and the implemented channel with the calculated percentage ratio (9.22) is the channel defined by the pattern (9.23).

$$\sigma_1 < \sigma_2 < \ldots < \sigma < \ldots < \sigma_{n-1} < \sigma_n => \begin{cases} |\sigma - \sigma_{i-1}| < |\sigma - \sigma_i| => \eta_i \uparrow \\ |\sigma - \sigma_{i-1}| > |\sigma - \sigma_i| => \eta_{i-1} \uparrow \end{cases} \tag{9.23}$$

Thus, a set with positive nuclear effective cross-sections can be found by applying a special operator (9.23) for such a general set, according to regularity (9.22), resulting in the most probable values of the nuclear effective cross-section.

$$U(\sigma_1, \sigma_2, \ldots, \sigma_{n-1}, \sigma_n) = (\sigma_i > 0) = \sigma_i, \ldots, \sigma_j, \ldots, \sigma_k \tag{9.24}$$

However, in this case, when the percentage ratio for the main reaction channel is determined, the remaining percentages are determined according to the ratios of the outputs of their channels. So, if the percentage ratio and the nuclear effective cross-section coincided with the first reaction channel, then the percentages of the following channels are determined, from where it is easy to calculate their nuclear effective cross-sections (9.25).

$$\begin{cases} \sigma_2 = -\dfrac{\ln\left(\frac{(1-\eta_1)Q_2}{\sum_{i=2}^{n} Q_i}\right)}{n_{A_{x_2}^{y_2}} dx_{A_{x_2}^{y_2}}} \\ \sigma_3 = -\dfrac{\ln\left(\frac{(1-\eta_1)Q_3}{\sum_{i=2}^{n} Q_i}\right)}{n_{A_{x_2}^{y_2}} dx_{A_{x_2}^{y_2}}} \\ \ldots \\ \sigma_n = -\dfrac{\ln\left(\frac{(1-\eta_n)Q_n}{\sum_{i=2}^{n} Q_i}\right)}{n_{A_{x_2}^{y_2}} dx_{A_{x_2}^{y_2}}} \end{cases} \tag{9.25}$$

In addition, the percentage ratio with the changes made is sufficient to represent new separating percentages with new values of the additional nuclear effective cross-sections of each of the channels, for the case under direct investigation, showing probabilistic reactions with their nuclear effective cross-sections (9.26).

$$a_{x_1}^{y_1} + A_{x_2}^{y_2} \rightarrow \begin{cases} B_{i(x_3)}^{i(y_3)} + b_{i(x_4)}^{i(y_4)} \\ \cdots \\ B_{j(x_3)}^{j(y_3)} + b_{j(x_4)}^{j(y_4)} \\ \cdots \\ B_{k(x_3)}^{k(y_3)} + b_{k(x_4)}^{k(y_4)} \end{cases} \leftrightarrow \begin{cases} \sigma_i \\ \cdots \\ \sigma_k \\ \cdots \\ \sigma_j \end{cases} \tag{9.26}$$

Thus, part of the analysis of the reaction initiation case is completed and after that it is necessary to pay attention to the further development of the reaction. So, after the reaction, most often, even among the probable channels, radioactive nuclei may be present, which must also be analyzed as decaying, in the form of their own long channels (9.27), most often with light escaping nuclei, as was done in (9.5).

$$B_{j(x_3)}^{j(y_3)} \rightarrow \begin{cases} C_{1(x_5)}^{1(y_5)} + c_{1(x_6)}^{1(y_6)} \\ C_{2(x_5)}^{2(y_5)} + c_{2(x_6)}^{2(y_6)} \\ \cdots \\ C_{n(x_5)}^{n(y_5)} + c_{n(x_6)}^{n(y_6)} \end{cases} \tag{9.27}$$

For each of the channels, the reaction output (9.28) is determined, which, due to the index, should not be mixed with (9.11), from where a new system of probabilistic nuclear reaction channels (9.29) is compiled using the same algorithm to determine the positive values of the outputs for each of the reaction channels.

$$Q_j = \left(M_{B_{j(x_3)}^{j(y_3)}} - M_{C_{i(x_5)}^{i(y_5)}} - M_{c_{i(x_6)}^{i(y_6)}} \right) c^2 \tag{9.28}$$

$$B_{j(x_3)}^{j(y_3)} \rightarrow \begin{cases} C_{i(x_5)}^{i(y_5)} + c_{i(x_6)}^{i(y_6)} \\ \cdots \\ C_{j(x_5)}^{j(y_5)} + c_{j(x_6)}^{j(y_6)} \\ \cdots \\ C_{k(x_5)}^{k(y_5)} + c_{k(x_6)}^{k(y_6)} \end{cases} \tag{9.29}$$

Most often, during this stage of analysis, only those channels remain, as a result of which stable nuclei and stable outgoing light particles or light nuclei remain, each of which will be produced. And to determine the percentage ratio, (9.13) is used for the present case, as a result of which it is possible to determine the percentage ratio, which will not be changed due to the absence of an incoming particle with the choice of new versions of the probability values of each of the channels of nuclear radioactive decay.

As a result of the transformations obtained and the results regarding the probabilistic channels of the nuclear reaction, according to the latest model (9.26) and the probabilistic channels of the reaction for the radioactive nucleus (9.29), the kinetic energies for the nuclei and outgoing particles in (9.30)–(9.37) can be calculated according to certain patterns.

$$T_{B^{i(y_3)}_{i(x_3)}} = \frac{\left(E_{a^{y_1}_{x_1}} - B_k + Q_i\right) M_{b^{i(y_4)}_{i(x_4)}}}{M_{B^{i(y_3)}_{i(x_3)}} + M_{b^{i(y_4)}_{i(x_4)}}} \tag{9.30}$$

$$T_{b^{i(y_4)}_{i(x_4)}} = \frac{\left(E_{a^{y_1}_{x_1}} - B_k + Q_i\right) M_{B^{i(y_3)}_{i(x_3)}}}{M_{B^{i(y_3)}_{i(x_3)}} + M_{b^{i(y_4)}_{i(x_4)}}} \tag{9.31}$$

$$T_{B^{j(y_3)}_{j(x_3)}} = \frac{\left(E_{a^{y_1}_{x_1}} - B_k + Q_i\right) M_{b^{j(y_4)}_{j(x_4)}}}{M_{B^{j(y_3)}_{j(x_3)}} + M_{b^{j(y_4)}_{j(x_4)}}} \tag{9.32}$$

$$T_{b^{j(y_4)}_{j(x_4)}} = \frac{\left(E_{a^{y_1}_{x_1}} - B_k + Q_i\right) M_{B^{j(y_3)}_{j(x_3)}}}{M_{B^{j(y_3)}_{j(x_3)}} + M_{b^{j(y_4)}_{j(x_4)}}} \tag{9.33}$$

$$T_{C^{i(y_5)}_{i(x_5)}} = \frac{\left(E_{B^{j(y_3)}_{j(x_3)}} + Q_j\right) M_{c^{i(y_6)}_{i(x_6)}}}{M_{C^{i(y_5)}_{i(x_5)}} + M_{c^{i(y_6)}_{i(x_6)}}} \tag{9.34}$$

$$T_{c^{i(y_6)}_{i(x_6)}} = \frac{\left(E_{B^{j(y_3)}_{j(x_3)}} + Q_j\right) M_{C^{i(y_5)}_{i(x_5)}}}{M_{C^{i(y_5)}_{i(x_5)}} + M_{c^{i(y_6)}_{i(x_6)}}} \tag{9.35}$$

$$T_{B^{k(y_3)}_{k(x_3)}} = \frac{\left(E_{a^{y_1}_{x_1}} - B_k + Q_i\right) M_{b^{k(y_4)}_{k(x_4)}}}{M_{B^{k(y_3)}_{k(x_3)}} + M_{b^{k(y_4)}_{k(x_4)}}} \tag{9.36}$$

$$T_{b^{k(y_4)}_{k(x_4)}} = \frac{\left(E_{a^{y_1}_{x_1}} - B_k + Q_i\right) M_{B^{k(y_3)}_{k(x_3)}}}{M_{B^{k(y_3)}_{k(x_3)}} + M_{b^{k(y_4)}_{k(x_4)}}} \tag{9.37}$$

After the kinetic energies for each of the nuclear reaction products have been determined, as probabilistic for the main channel and probabilistic for the decay channels of radioactive nuclei, it is necessary to present the distribution according to the currents of incoming beams and products of each of the reaction channels. To do this, it is initially assumed to determine the previously calculated beam cross-section (9.38), from which it is possible to deduce from the formula for the beam current the amount of incoming particles (9.39), which was not previously determined during the calculation of percentages and the nuclear effective cross-section for the analyzed case.

$$S_{a_{x_1}^{y_1}} = \pi r^2_{a_{x_1}^{y_1}} \tag{9.38}$$

$$I_{a_{x_1}^{y_1}} = N_0 e S_{a_{x_1}^{y_1}} v_{a_{x_1}^{y_1}} \Rightarrow N_0 = \frac{I_{a_{x_1}^{y_1}}}{e S_{a_{x_1}^{y_1}} v_{a_{x_1}^{y_1}}} \tag{9.39}$$

During the analysis of the reaction situation, the incoming beams have a certain amount of current, set initially and most often limited by the capabilities of the accelerator system, from where it becomes possible to determine the number of incoming particles (9.39) and from where it is possible to determine the total charge of the incoming beam (9.40).

$$q_{a_{x_1}^{y_1}} = N_0 e \tag{9.40}$$

Finally, after the percentage distributions have been made, the probable channels with their nuclear effective cross-sections, the kinetic energies of the reaction products and the number of incoming particles have been determined, it is possible to fully present a model of the distribution of incoming particles in a variety of all channels, including in a variety of probabilistic channels of radioactive decay, from where it is possible to determine according to previously determined kinetic energies reaction products, exact quantitative ratios (9.41).

$$a_{x_1}^{y_1} + A_{x_2}^{y_2} \rightarrow \begin{cases} B_{i(x_3)}^{i(y_3)} - E_{B_{i(x_3)}^{i(y_3)}} - n\left(B_{i(x_3)}^{i(y_3)}\right) = N_0 e^{-n_{A_{x_2}^{y_2}} \sigma_i dx_{A_{x_2}^{y_2}}} \\ b_{i(x_4)}^{i(y_4)} - E_{b_{i(x_4)}^{i(y_4)}} - n\left(b_{i(x_4)}^{i(y_4)}\right) = N_0 e^{-n_{A_{x_2}^{y_2}} \sigma_i dx_{A_{x_2}^{y_2}}} \\ \ldots \\ B_{j(x_3)}^{j(y_3)} \rightarrow \begin{cases} C_{i(x_5)}^{i(y_5)} - E_{C_{i(x_5)}^{i(y_5)}} - n\left(C_{i(x_5)}^{i(y_5)}\right) = n\left(B_{j(x_3)}^{j(y_3)}\right) * 2^{-\frac{t}{T}} \\ c_{i(x_6)}^{i(y_6)} - E_{c_{i(x_6)}^{i(y_6)}} - n\left(c_{i(x_6)}^{i(y_6)}\right) = n\left(B_{j(x_3)}^{j(y_3)}\right) * 2^{-\frac{t}{T}} \\ \ldots \\ C_{j(x_5)}^{j(y_5)} - E_{C_{j(x_5)}^{j(y_5)}} - n\left(C_{j(x_5)}^{j(y_5)}\right) = n\left(B_{j(x_3)}^{j(y_3)}\right) * 2^{-\frac{t}{T}} \\ c_{j(x_6)}^{j(y_6)} - E_{c_{j(x_6)}^{j(y_6)}} - n\left(c_{j(x_6)}^{j(y_6)}\right) = n\left(B_{j(x_3)}^{j(y_3)}\right) * 2^{-\frac{t}{T}} \\ \ldots \\ C_{k(x_5)}^{k(y_5)} - E_{C_{k(x_5)}^{k(y_5)}} - n\left(C_{k(x_5)}^{k(y_5)}\right) = n\left(B_{j(x_3)}^{j(y_3)}\right) * 2^{-\frac{t}{T}} \\ c_{k(x_6)}^{k(y_6)} - E_{c_{k(x_6)}^{k(y_6)}} - n\left(c_{k(x_6)}^{k(y_6)}\right) = n\left(B_{j(x_3)}^{j(y_3)}\right) * 2^{-\frac{t}{T}} \end{cases} \\ b_{j(x_4)}^{j(y_4)} - E_{b_{j(x_4)}^{j(y_4)}} - n\left(b_{j(x_4)}^{j(y_4)}\right) = N_0 e^{-n_{A_{x_2}^{y_2}} \sigma_j dx_{A_{x_2}^{y_2}}} \\ \ldots \\ B_{k(x_3)}^{k(y_3)} - E_{B_{k(x_3)}^{k(y_3)}} - n\left(B_{k(x_3)}^{k(y_3)}\right) = N_0 e^{-n_{A_{x_2}^{y_2}} \sigma_k dx_{A_{x_2}^{y_2}}} \\ b_{k(x_4)}^{k(y_4)} - E_{b_{k(x_4)}^{k(y_4)}} - n\left(b_{k(x_4)}^{k(y_4)}\right) = N_0 e^{-n_{A_{x_2}^{y_2}} \sigma_k dx_{A_{x_2}^{y_2}}} \end{cases} \tag{9.41}$$

Thus, after analyzing the quantitative ratio, it becomes obvious that the elementary particles will simply leave the resulting split core, where the droplet model of nuclear reactions is most often used. The parameters of the outgoing particles, along with their number, kinetic energies, and beam cross sections, which are partly equal to (9.38), are obvious.

It remains to analyze the situation with the remaining cores. One of the main indicators required for determination is the variable temperature of the formed nuclei as a result of the nuclear reaction, which can be determined from the total energy of the components, which can be represented as the products of the energies of each of the nuclei together with the number of nuclei formed, resulting in the total energy of the nuclei of each of the nuclear reaction channels (9.42).

$$E = \begin{cases} Q_i = E_{B^{i(y_3)}_{i(x_3)}} n\left(B^{i(y_3)}_{i(x_3)}\right) \\ \cdots \\ Q_j = E_{C^{i(y_5)}_{i(x_5)}} n\left(C^{i(y_5)}_{i(x_5)}\right) \\ \cdots \\ Q_k = E_{C^{k(y_5)}_{k(x_5)}} n\left(C^{k(y_5)}_{k(x_5)}\right) \end{cases} \tag{9.42}$$

Since each of the reaction channels is carried out in a certain percentage during the reaction, from this it is possible to determine the total energy as the sum of the energies of the forming nuclei (9.43), in cases when particles fly out of them and do not fly out, which may be useful for determining the temperature in (9.47), for which it is necessary to determine the quantitative ratio with the specific heat capacity of the plate to its mass, for which its volume (9.44) is determined, and from there the initial mass (9.45) is determined from the density value, However, since the plate consumed part of the nuclei to carry out the nuclear reaction, it is necessary to determine the resulting mass (9.46).

$$Q = \sum_n E_{i,j,k} \tag{9.43}$$

$$V_{A^{y_1}_{x_1}} = S_{A^{y_1}_{x_1}} dx_{A^{y_1}_{x_1}} \tag{9.44}$$

$$m'_{A^{y_1}_{x_1}} = V_{A^{y_1}_{x_1}} \rho_{A^{y_1}_{x_1}} \tag{9.45}$$

$$m_{A^{y_1}_{x_1}} = \left(\frac{m'_{A^{y_1}_{x_1}}}{M_{A^{y_1}_{x_1}}} - N_0\right) M_{A^{y_1}_{x_1}} \tag{9.46}$$

$$Q = cm\Delta T \Rightarrow \Delta T = \frac{Q}{c_{A^{y_1}_{x_1}} m_{A^{y_1}_{x_1}} + \sum_{i,j,k}^{i} c_{B^{i(y_3)}_{i(x_3)}} m_{B^{i(y_3)}_{i(x_3)}}} \tag{9.47}$$

Thus, as a result of the reaction, the necessary particular characteristic for the nuclei has been determined, along with their number, mass, and temperature, however, we will have to return to this issue, since this is precisely the internal temperature, but it is worth paying attention to the outgoing particles now. Each of the outgoing particles in each of the reaction channels has its own total charge for each of the beams (9.48) with its own velocities (9.49) and currents (9.50).

$$q_{b^{i(y_4)}_{i(x_4)}} = Z_{b^{i(y_4)}_{i(x_4)}} n_{b^{i(y_4)}_{i(x_4)}} e \tag{9.48}$$

$$v_{b_{i(x_4)}^{i(y_4)}} = c\sqrt{1 - \left(\frac{1}{1 + \frac{E_{b_{i(x_4)}^{i(y_4)}}}{m_{b_{i(x_4)}^{i(y_4)}} c^2}}\right)^2} \tag{9.49}$$

$$I_{b_{i(x_4)}^{i(y_4)}} = n_{b_{i(x_4)}^{i(y_4)}} eS_{a_{x_1}^{y_1}} v_{b_{i(x_4)}^{i(y_4)}} \tag{9.50}$$

Before starting the reaction, the input Coulomb barrier was determined, after which various internal indicators began. Now that each of them is known, it remains to determine the output results, which can be represented as an outgoing Coulomb barrier (9.52), with the radii of each of the formed nuclei (9.51).

$$R_{B_{i(x_3)}^{i(y_3)}} = 1.4 * (i(y_3))^{\frac{1}{3}} \tag{9.51}$$

$$B'_{k\left(b_{i(x_4)}^{i(y_4)}\right)} = \frac{\left|q_{b_{i(x_4)}^{i(y_4)}}\right|\left|q_{B_{i(x_3)}^{i(y_3)}}\right|}{R_{B_{i(x_3)}^{i(y_3)}}} \frac{e^2}{\hbar c} \hbar c \tag{9.52}$$

It should be noted here that the outgoing Coulomb barrier, according to experimental data, almost always acts in cases when the outgoing particle has a positive charge, exactly like the nucleus and quite rarely in cases with negative charges. So, when the nucleus and the outgoing particle have the same charge, the kinetic energies add up, in the opposite cases they should reach each other, however, according to the resulting work, it is necessary to postulate that in most cases, such an addition is extremely rare for a charge of different names.

Along with the above, it is also impossible to say that there are no nuclear reactions carried out in which there would be no energy negation of the outgoing particle, which may be one of the topics of further research. However, one should not discount the possibility of an explanation using existing nuclear models, for example, a drop or shell model of the nucleus. The first described case with charges of the same name for the energies of outgoing particles with an outgoing Coulomb barrier can be defined as the sum in (9.53).

$$E'_{b_{i(x_4)}^{i(y_4)}} = E_{b_{i(x_4)}^{i(y_4)}} + B'_{k\left(b_{i(x_4)}^{i(y_4)}\right)} \tag{9.53}$$

From the energy expression obtained, it is possible to determine the velocity of the outgoing particle (9.54) with their own currents (9.55).

$$v'_{b^{i(y_4)}_{i(x_4)}} = c\sqrt{1 - \left(\frac{1}{1 + \frac{E'_{b^{i(y_4)}_{i(x_4)}}}{m_{b^{i(y_4)}_{i(x_4)}} c^2}}\right)^2} \tag{9.54}$$

$$I'_{b^{i(y_4)}_{i(x_4)}} = n_{b^{i(y_4)}_{i(x_4)}} e S_{a^{y_1}_{x_1}} v'_{b^{i(y_4)}_{i(x_4)}} \tag{9.55}$$

Now, it only remains to return to the issue of the energies of the formed nuclei, which also have changed energies (9.56), from where we can calculate the value for the changed total energy (9.57), and therefore the temperature (9.58).

$$E'_{B^{i(y_4)}_{i(x_4)}} = E_{B^{i(y_4)}_{i(x_4)}} + B'_{k\left(b^{i(y_4)}_{i(x_4)}\right)} \tag{9.56}$$

$$Q' = \sum_n E'_{i,j,k} \tag{9.57}$$

$$\Delta T' = \frac{Q'}{c_{A^{y_1}_{x_1}} m_{A^{y_1}_{x_1}} + \sum^{i}_{i,j,k} c_{B^{i(y_3)}_{i(x_3)}} m_{B^{i(y_3)}_{i(x_3)}}} \tag{9.58}$$

As a result, all the necessary results that needed to be determined were identified. Thus, we can say that the nuclear reaction (9.1) with all its reaction channels has been fully analyzed.

9.4 Results

As a result of the analysis, the following conclusions are drawn:

- The condition of the task: a nuclear reaction of the form (2.1) was investigated with initially specified parameters in the form of the kinetic energy of the directed beam and the mass of all participating particles in the nuclear reaction in a. e. m.;
- During the Rutherford scattering analysis, strictly defined results were achieved in the form of the value of the nuclear effective cross section and the maximum distance convergence at critical velocity with the nucleus and the incoming particle;
- Directed beams consume a certain amount of energy to overcome the Coulomb barrier, having residual energy—it's found value is indicated;
- A list of the outgoing particles from the main channel of the nuclear reaction is made, including various groups of gamma rays, if any, indicating the kinetic energy (where, if necessary, their classification by energy is also compiled), the charge and current of each of them;

- If the formed particles can probabilistically interact (similar to the annihilation of positrons and electrons), then this is indicated and an additional list with all relevant accounts is provided;
- In accordance with each reaction, a list is provided with all the work performed in Joules and capacities in Watts for each type of radiation with all the components, along with the number of nuclei formed, their masses, other additional results and energy values;
- The general conclusions on the objectives of this study are indicated—the general study/generation of electrical energy/establishment of conclusions on some precise aspect, etc., as well as conclusions in the relevant area: the amount of energy generated, conclusions on the necessary aspect, general scientific data, conclusions, etc.

Thus, based on mathematical analysis, the energies of the formed particles and their nature of origin are calculated. Nuclear reactions with bombarding charged particles with high and low kinetic energies of target nuclei, as well as the resonance state of these nuclear reactions, have been studied. The present method and algorithm is currently new with the results coming out.

9.5 Conclusion

Based on the theoretical calculations carried out, it is indicated that using the proposed analysis methods it is possible to obtain certain results that correlate with experimental data with a small error. At the same time, the use of the obtained model makes it possible to simulate the implementation of various types of nuclear reactions with bringing to a state of resonance. At the same time, the best variations can be selected among them, preserving their best efficiency, while maintaining maximum currents and energies of the outgoing beams, capable of further use through MHD generation. Thus, the proposed technique is a new approach to studying the nuclear interactions of high-energy particles with stationary target nuclei, considering the full-fledged effectiveness of each of the nuclear reaction channels, their own resultant parameters for each of the reaction products, and other data.

References

1. I.X. Aliev, F.M. Sharofutdinov, *The Use of Accelerators and Phenomena of Collision of Elementary Particles with High-Order Energy to Generate Electrical Energy. The "Electron" Project. Monograph* (Electronic Laboratory, The Scientific School "Electron", Publishing Solutions. "Ridero", Margilan, 2021)
2. I.X. Aliev, *New Parameters for Nuclear Reactions to be Carried Out at an LCU-EPD-300 Type Charged Particle Accelerator. The "Electron" Project. Monograph* (Electron Laboratory, The Scientific School "Electron", Publishing Solutions "Ridero", Margilan, 2022)
3. I.X. Aliev, Aluminum resonant nuclear reaction. Int. Sci. J. **3**(3), 24–44 (2022)

4. E. Rutherford, On the retardation of the α particle from radium in passing through matter. Philos. Mag. **12**(68), 134–146 (1906). https://doi.org/10.1080/14786440609463525
5. L. Coraggio, G. De Gregorio, T. Fukui, A. Gargano, Y.Z. Ma, Z.H. Cheng, F.R. Xu, The role of three-nucleon potentials within the shell model: past and present. Prog. Part. Nucl. Phys. **134**, 104079 (2024). https://doi.org/10.1016/j.ppnp.2023.104079
6. H. Geiger, On the scattering of α-particles by matter. Proc. Royal Soc. A **81**(546), 174–177 (1908). https://doi.org/10.1098/rspa.1908.0067
7. H. Geiger, E. Marsden, On the diffuse reflection of the α-particles. Proc. Royal Soc. A **82**(557), 495 (1909). https://doi.org/10.1098/rspa.1909.0054
8. H. Geiger, The scattering of the α-particles by matter. Proc. Royal Soc. A **83**(565), 492–504 (1910). https://doi.org/10.1098/rspa.1910.0038
9. A. Sorensen, K. Agarwal, K.W. Brown, Z. Chajęcki, Dense nuclear matter equation of state from heavy-ion collisions. Prog. Part. Nucl. Phys. **134**, 104080 (2024). https://doi.org/10.1016/j.ppnp.2023.104080
10. D. Stefan, Experimental exploration of the 3D nucleon structure. Prog. Part. Nucl. Phys. **133**, 104069 (2023). https://doi.org/10.1016/j.ppnp.2023.104069
11. C. Vuille, J. Ipser, J. Gallagher, Einstein-Proca model, micro black holes, and naked singularities. Gen. Relativ. Gravit. **34**(5), 689 (2002). https://doi.org/10.1023/a:1015942229041
12. M.G. Mayer, On closed shells in nuclei. II. Phys. Rev. **75**(12), 1969–1970 (1949). https://doi.org/10.1103/PhysRev.75.1969
13. G. Aarts, J. Aichelin, C. Allton, A. Athenodorou, Phase transitions in particle physics: results and perspectives from lattice quantum chromo-dynamics. Prog. Part. Nucl. Phys. (2023). https://doi.org/10.1016/j.ppnp.2023.104070
14. A.S. Eddington, On the radiative equilibrium of the stars. Mon. Not. R. Astron. Soc. **77**, 16–35 (1916). https://doi.org/10.1093/mnras/77.1.16
15. A. Cho, Aging atom smasher runs all out in race for most coveted particle. Science **312**(5778), 1302–1303 (2006). https://doi.org/10.1126/science.312.5778.1302
16. T. Feder, Accelerator school travels university circuit. Phys. Today **63**(2), 20–22 (2010). https://doi.org/10.1063/1.3326981
17. J. Ullrich, A. Rudenko, R. Moshammer, Free-electron lasers: new avenues in molecular physics and photochemistry. Annu. Rev. Phys. Chem. **63**(1), 635–660 (2012). https://doi.org/10.1146/annurev-physchem-032511-143720
18. A. Mak, G. Shamuilov, P. Salén, D. Dunning, J. Hebling, Y. Kida, R. Kinjo, B.W. McNeil, T. Tanaka, N. Thompson, Z. Tibai, Attosecond single-cycle undulator light: a review. Rep. Prog. Phys. **82**(2), 025901 (2019). https://doi.org/10.1088/1361-6633/aafa35
19. R.J. Van de Graaff, K.T. Compton, L.C. Van Atta, The electrostatic production of high voltage for nuclear investigations. Phys. Rev. **43**(3), 149–157 (1933). https://doi.org/10.1103/PhysRev.43.149
20. L. Cassiday, Hair-raising technique detects drugs, explosives on human body. Science (2014). https://doi.org/10.1126/article.22861
21. M. Pfützner, I. Mukha, S.M. Wang, Two-proton emission and related phenomena. Prog. Part. Nucl. Phys. **132**, 104050 (2023). https://doi.org/10.1016/j.ppnp.2023.104050
22. S. Cao, G.-Y. Qin, Medium response and jet–hadron correlations in relativistic heavy-ion collisions. Annu. Rev. Nucl. Part. Sci. **73**, 205–229 (2023). https://doi.org/10.1146/annurev-nucl-112822-031317
23. M. Monthioux, V.L. Kuznetsov, Who should be given the credit for the discovery of carbon nanotubes? Carbon **44**(9), 1621–1623 (2006). https://doi.org/10.1016/j.carbon.2006.03.019
24. M. Kokarneswaran, P. Selvaraj, T. Ashokan, S. Perumal, P. Sellappan, K.D. Murugan, Discovery of carbon nanotubes in sixth century BC potteries from Keeladi, India. Sci. Rep. **10**(1), 19786 (2020). https://doi.org/10.1038/s41598-020-76720-z
25. A.C. Torres-Dias, From mesoscale to nanoscale mechanics in single-wall carbon nanotubes. Carbon **123**, 145–150 (2017). https://doi.org/10.1016/j.carbon.2017.07.036

Chapter 10
Some Questions of Thermonuclear Fusion Physics: The New Mechanism of the ^2H-^3H Reaction

V. S. Okunev

Abstract The article states that the mechanism of the fusion reaction of deuterium and tritium nuclei is not completely clear. The overwhelming majority of modern thermonuclear reactors and their projects are focused on the use of deuterium–tritium plasma. It has been established that the mechanism of fusion of deuterium and tritium nuclei differs from the classical mechanism of a compound nucleus proposed by N. Bohr and from the direct interaction proposed by S. T. Butler. The article proposes a new mechanism that assumes the formation of a compound nucleus of helium-5. Unlike the Bohr mechanism, in the proposed model the compound nucleus ^5He formally refers to a non-existent one. Its average lifetime is about 7×10^{-22} s and is comparable with the characteristic time of strong interaction ($\tau_S \sim 10^{-23} \ldots 10^{-22}$ s). The reaction rate is typical for direct nuclear interactions, and the sequence of events, the mechanism and the dependence of the cross section on the energy of the interacting nuclei are typical for nuclear reactions realized through the formation of a compound nucleus. Thus, a new mechanism of the fusion reaction has been revealed. An overview of thermonuclear reactors is presented.

10.1 Introduction

10.1.1 Background

The projected growth of the world population, the desire to improve the quality of life, and the development of information technology (associated with the growth of energy consumption by server equipment) require an increase in energy production. The main requirements for energy sources of the future are economic efficiency (at

V. S. Okunev (✉)
Bauman Moscow State Technical University, Moscow, Russia
e-mail: okunevvs@bmstu.ru

© The Author(s), under exclusive license to Springer Nature Switzerland AG 2025

A. Gibadullin et al. (eds.), *Proceedings of the 4th International Scientific and Practical Symposium on Materials Science and Physics*, Springer Proceedings in Physics 318,
https://doi.org/10.1007/978-3-031-94805-3_10

least the production of useful energy); safety in the broad sense of the word; long-term fuel supply (ideally, self-sufficiency in fuel); and the ability to produce energy on the necessary scale.

Fission reactors of heavy nuclei, fusion reactors of light nuclei and electronuclear installations (sub-critical accelerator driven systems) can lay claim to the role of energy sources of the future. The operation of such systems is based on nuclear fission and fusion reactions. Energy can be obtained by dividing heavy nuclei into lighter fragments and synthesizing light nuclei into heavier ones. In this case, the greatest energy is released if the reaction products are magic or (better) doubly magic nuclei. For these nuclei, the specific binding energy is maximal.

The technologies of fission reactors of heavy nuclei are well developed and brought to the "engineering" level. However, such reactors represent a great potential danger to the population and the environment. From the very beginning of the development of fission reactor technologies, researchers have encountered accidents of varying severity. Thus, the first major radiation accidents occurred during the development of nuclear materials for the first atomic bombs, based on the fission reaction of heavy nuclei.

The first heavy nuclear fission reactor (plutonium-239), Chicago Pile-1, was built in 1942 at the University of Chicago under the direction of Enrico Fermi [1, 2]. In 1944, in the United States (Oak Ridge National Laboratory, Tennessee), an explosion of uranium hexafluoride occurred during an attempt to clean a pipe in a laboratory uranium enrichment device. In the USSR, the first serious radiation accident occurred in 1948 at the "Mayak" plant (Chelyabinsk region) the very next day after the nuclear fission reactor was brought to capacity to produce weapons-grade plutonium. In 1952, in Canada, due to a technical error by personnel, the first radiation accident occurred at a nuclear power plant (NPP) with a research reactor NRX (National Research Experimental) at Chalk River Laboratories (Ontario).

In total, from the start of operation of NPPs until the end of 2007, more than 150 incidents and accidents of varying degrees of complexity occurred in 14 countries around the world [3]. The most severe of them were the accidents at the Windscale-1 reactor (1957, Cumbria County in northwest England), the 2nd unit of the Three Mile Island NPP (1979, Pennsylvania, USA), the 4th unit of the Chernobyl NPP (1986, Pripyat, Kyiv region, Ukraine, USSR), and at four units of the Fukushima-1 NPP (2011, Fukushima Prefecture, Japan).

Formally, for the practical implementation of heavy nuclear fission reactors in the problems of their optimal design, it is necessary to satisfy a mathematical constraint of the equality type for the effective coefficient k of neutron multiplication (criticality condition: $k = 1$). For the implementation of fusion reactors, it is necessary to satisfy a constraint of the inequality type (Lawson criterion: $N \tau > L$, where N is the concentration of nuclei; τ is the plasma confinement time; L is the Lawson number). From the point of view of the practical implementation of a technical object, the constraint of the inequality type is much easier to fulfill.

The paradox is that nuclear fusion reactors are not yet available, while fission reactors are widely used. By the end of 2024, there were 415 nuclear power units in operation worldwide with a total installed electrical capacity of about 373.735

GW, and 63 nuclear power units with an electrical capacity of 66.1 GW were under construction [4]. The world's nuclear power industry with fission reactors generated 20,053 reactor-years [4].

For several decades, the problem of implementing controlled thermonuclear fusion has been one of the main problems of physics, representing the greatest interest [5, 6]. In his Nobel lecture, Academician V. L. Ginsburg puts it in first place among the physical problems of the beginning of the twenty-first century [5].

In recent decades, the task of practical implementation of thermonuclear fusion installations for energy needs has come to the fore, which requires the development of thermonuclear technologies. From the point of view of obtaining energy, the mechanism of a specific thermonuclear reaction is of no interest. It is important that this is a nuclear reaction, i.e. a reaction realized due to nuclear (as a manifestation of strong) interaction (in this case, its cross section is quite large); and that this reaction occurs with the release of energy.

For over 30 years, experimental research has been actively conducted on the use of a mixture of deuterium (^2H, or D, or deuteron) and tritium (^3H, or T, or triton) nuclei as fuel for thermonuclear reactors.

Rutherford was the first to observe a nuclear reaction in 1919, bombarding nitrogen atom nuclei with α-particles (^4He nuclei) [7]:

$$^{14}N + {}^{4}He \rightarrow {}^{17}O + p.$$

In this reaction, the proton p was discovered (as it turned out later, the ^1H nucleus). In this and other reactions (α, p), the presence of resonances in the dependence of the reaction cross section on the kinetic energy of the α-particles was discovered. This indicates that the reactions are realized through a compound nucleus. A theoretical model of a nuclear reaction proceeding through the formation of a compound nucleus was proposed by the Danish theoretical physicist N. Bohr in 1936 [8]. In essence, this reaction can be considered a synthesis reaction of light nuclei $^{14}N + {}^{4}He$, in which a heavier nucleus ^{17}O is formed. The reaction is realized at α-particle energy of more than 1.45 MeV. The reaction energy is 1.18 MeV:

The first fusion reaction of light nuclei to obtain energy was carried out by J. Cockforth and E. Walton (Nobel Prize in Physics 1951) at the Cavendish Laboratory (Cambridge, UK) in 1932 at a proton accelerator [9]:

$$p + {}^{7}Li\ (\text{target}) \rightarrow 2^{4}He + \gamma + 17.3\text{M}\ni\text{B}.$$

Later, other fusion reactions were realized in laboratories around the world. Among them is the D-D reaction, which has two approximately equally probable channels:

$$^{2}H + {}^{2}H \rightarrow {}^{3}He + n + 3.25\,\text{MeV};$$
$$^{2}H + {}^{2}H \rightarrow {}^{3}H + p + 4.0\,\text{MeV}.$$

D-T reaction

$$^{2}\mathrm{H} + {}^{3}\mathrm{H} \rightarrow {}^{4}\mathrm{He} + \mathrm{n} + 17.6\,\mathrm{MeV};$$

and also reactions

$$^{3}\mathrm{He} + {}^{2}\mathrm{H} \rightarrow {}^{4}\mathrm{He} + \mathrm{p} + 18.3\,\mathrm{MeV};$$
$$^{6}\mathrm{Li} + {}^{2}\mathrm{H} \rightarrow 2^{4}\mathrm{He} + 22.4\,\mathrm{MeV}.$$

In 1970, the International Council on Thermonuclear Research was formed under the auspices of the IAEA. Research on the problem of controlled thermonuclear fusion, which has been conducted in our country and abroad in recent decades, concerns mainly the technologies of thermonuclear reactors. Research in recent years has been devoted to the technologies, materials of thermonuclear reactors, physics and diagnostics of plasma, mathematical modeling of transport processes in plasma (see [10–12], as well as the Proceedings of the annual International Conferences on Plasma Physics and Controlled Thermonuclear Fusion, held in Zvenigorod).

Work is underway on the ITER project [13, 14]. In 1991, the European JET facility produced D-T plasma with a low tritium content for the first time [15]. In 1993, the TFTR reactor in the USA conducted the first experiment using 50% D–50% T plasma [16]. However, the mechanism of the D-T fusion reaction itself is of no practical interest.

The physics of thermonuclear fusion and thermonuclear reactions is considered a solved problem [17]. But is it really solved?

10.1.2 Problem Statement

The key issue in plasma physics and thermonuclear fusion is the question of the mechanisms of nuclear fusion reactions. Despite significant advances in the development of thermonuclear technologies, this issue has not yet been closed.

According to the mechanism of interaction, nuclear reactions (all inelastic processes) are divided into two types:

- Reactions are realized with the formation of a compound nucleus. The mechanism was proposed by N. Bohr in 1936. This process is realized in two stages. It occurs at a not very high kinetic energy of the colliding particles (approximately up to ~ 10 MeV).
- The reactions are realized as direct nuclear reactions that occur during the nuclear time required for the particle to cross the nucleus. Direct interactions were discovered by S. Butler in 1953. This mechanism mainly manifests itself at high energies of the bombarding particles.

It is generally accepted that deuterium nuclei can interact with other atomic nuclei both to form a compound nucleus and through direct nuclear interaction. However, there is reason to believe that the mechanism of the D-T reaction at characteristic temperatures of thermonuclear plasma is different [18].

10.2 Materials and Methods

The research methods are based on a comprehensive analysis of information from existing nuclear databases and information published in numerous articles. The article uses the following databases: ROSFOND [19], NUBASE2016 [20], NUBASE2020 [21] and NNDF/B-VII.1 [22].

10.2.1 Fuel Materials for Thermonuclear Fusion Reactors

All light elements from hydrogen isotopes to lithium can be considered as potential fuel materials for thermonuclear fusion reactors. Several reactions of light nuclei synthesis are known. The cross sections of fusion reactions involving lithium nuclei are relatively small and require high energy to implement. The $^{2}H + ^{3}He$ reaction is attractive because it does not produce neutrons. Neutrons can only occur in accompanying D-D reactions. As a result, the neutron yield decreases by 20–50 times compared to the D-T reaction, which leads to a decrease in radioactive waste [9].

Charged particles—products of the fusion reaction—have insufficient energy to overcome the Coulomb barrier with possible subsequent activation of materials. Nuclides ^{3}H and ^{3}He are practically absent in nature. They are obtained artificially. It is also proposed to deliver ^{3}He to Earth from the Moon: ^{3}He is a product of the interaction of the solar wind with lunar soil. The surface layer of the Moon is saturated with this helium isotope to a depth of 3…5 m [9].

The cross section of the D-T reaction and its rate are hundreds of times higher than the cross section σ and rate of the D-D reaction (Fig. 10.1). Consequently, it is much easier to provide conditions for the production of useful energy for the deuterium and tritium fusion reaction. Fusion reactions involving nuclei heavier than helium or tritium (lithium, boron) are characterized by significantly smaller reaction cross sections and rates of their occurrence and reach noticeable values only for energies of the order of 100 keV. Achieving such energies (temperatures) in thermonuclear installations is currently not possible, so reactions of hydrogen isotope fusion may have practical application in the near future.

The use of accelerator technology to obtain energy in fusion reactions is inefficient: the energy obtained is much less than the energy expended. The cross section of the D-T reaction reaches its maximum value (about 5 b) when the interacting particles have a relative approach energy of about 200 keV (see Fig. 10.1). At energy of 20 keV, the cross section becomes less than 0.1 b. This means that out of a million accelerated

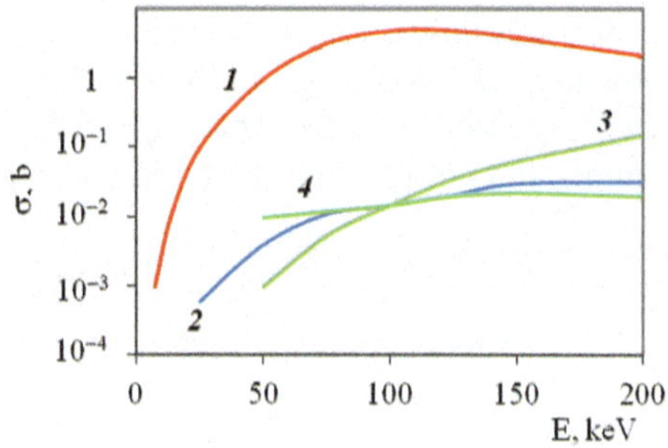

Fig. 10.1 Dependence of micro cross-sections σ of reactions of fusion of light nuclei on the energy of their relative approach (1b /barn/ = 10^{-24} cm^2) [23]: 1—D + T → ^{4}He + n; 2—D + D → ^{3}He + n; 3—D + ^{3}He → ^{4}He + p; 4—T + T → ^{4}He + 2n

particles hitting the target, no more than one enters into a nuclear interaction. The rest dissipate their energy on the electrons of the target atoms and slow down to speeds at which the reaction becomes impossible. Therefore, the greatest success in producing useful energy should be expected in thermonuclear fusion reactions.

Fusion reactors are designed to use light nuclei—mainly heavy hydrogen isotopes. The ^{2}H + ^{3}He reaction (neutron-free thermonuclear fusion) is attractive, but not yet justified. Other neutron-free fusion reactions are known: ^{2}H + ^{6}Li; ^{1}H + ^{6}Li; ^{3}He + ^{6}Li; ^{3}He + ^{3}He; ^{1}H + ^{7}Li; ^{1}H + ^{11}B.

The fuel base of D-T fusion reactors is quite broad. Deuterium is found in seawater: its reserves are generally available and very large (the world ocean covers 71% of the Earth, and deuterium accounts for about 0.016% of the total number of hydrogen atoms in water). In addition, a fusion reactor contains much less radioactive material than a fission reactor. As a result, the consequences of an accidental release of radioactive materials are less dangerous. In reactions of fusion of light nuclei, fewer long-lived radioactive nuclides are formed. Controlled fusion allows for the direct generation of electricity.

For the fusion reaction of nuclei (positively charged particles) to occur, the interacting particles must have energy sufficient to overcome the Coulomb barrier, the value of which is proportional to the product of their charges. There are several ways to overcome the Coulomb barrier and bring the nuclei closer to the distance r_S of the nuclear forces, when nuclear attraction dominates over Coulomb repulsion.

Heating matter (nuclei) to high temperatures (~ 10^8 K and higher) is the basis of thermonuclear fusion. The use of accelerator technology to accelerate nuclei to energies that allow overcoming Coulomb repulsion is the basis of the operation of neutron generators in reactions of fusion of light nuclei.

The introduction of a muon (μ) into the orbit of an atom that is part of a hydrogen molecule (muon catalysis) leads to the convergence of nuclei to the radius of action of nuclear forces (a muon is significantly heavier than an electron). Fusion reactions are possible, realized due to the tunnel effect. The stimulus for the development of cold nuclear fusion research was an interview with M. Fleshman and S. Pons (USA) on March 23, 1989 in the newspaper "Financial Times". They claimed that it was possible to create an industrial-scale energy source based on the fusion of heavy hydrogen nuclei at room temperature. However, the effect was not confirmed, and cold nuclear fusion was deemed unsuitable for industrial-scale energy production.

10.2.2 Thermonuclear Reactors. "Inherent Safety"

In 1946, British physicists J. P. Thomson and M. Blackman applied for a patent on a toroidal device for a thermonuclear reaction in a gas as a result of electromagnetic radiation pressure [24]. In the USSR in 1950, O. A. Lavrentiev formulated the problem of controlled thermonuclear fusion and proposed a constructive solution for it [25]. Lavrentyev did not have a higher education and in 1950 he entered the physics department of Moscow State University. In 1951, the American physicist L. Spitzer proposed a system for holding hot plasma with a magnetic field—a stellarator [26]. The first thermonuclear reactor T-1 (tokamak) was launched in 1954 (Kurchatov Institute, Moscow, Russia, USSR). In total, about 300 tokamak-type reactors were built worldwide.

Unlike nuclear fission reactors, in thermonuclear fusion reactors with both inertial and magnetic plasma confinement it is much easier to eliminate accidents with unacceptable emissions of radioactive substances. For such reactors, the ideals of internal self-protection are achievable. The most developed technologies are those of the "tokamak" type of reactors with magnetic plasma confinement. This is the simplest plasma reactor to implement.

Compared with fission reactors, fusion reactors will be characterized by significantly lower emissions of radioactive substances (including during accidents); no need for preliminary chemical treatment of waste before fabrication or disposal; the possibility of accident-free storage of waste on the territory of the thermonuclear power station (the reprocessing plant can be integrated on a single site with the thermonuclear power station); the possibility of disposal of part of the spent materials in near-surface formations without the need for careful monitoring of disposal sites to ensure the safety of future generations. In the event of any accidents at a thermonuclear facility that are not deterministically excluded, there is no need to evacuate the population living near the station.

In 1978, British chemical engineer Trevor Asher Kletz proposed a concept that involves reducing rather than controlling the hazards of chemical and process industry facilities [27]. Later in 1984, Kletz introduced the term "inherent safety" [28]. Following the accident at the Three Mile Island NPP (1979), in 1985, employees of the Oak Ridge Associated Universities Institute for Energy Analysis (Oak Ridge, Tennessee, U.S.) Irving Spiewak and Alvin Martin Weinberg (Director of the Institute for Energy Analysis) introduced the term "inherent safety" in relation to nuclear fission reactors [29].

The internal safety of reactors with magnetic plasma confinement is mainly related to instability, disruption or termination of plasma burning in any emergency situations. In this case, the reactor stops working. The types of emergency modes traditionally considered for nuclear fission reactors (uncontrolled increase in power and violation of heat removal conditions) cannot lead to an accident in thermonuclear reactors. The first event is excluded deterministically; the second is safe due to the small residual energy release after the reactor is stopped. Magnetic systems are characterized by the greatest stored energy, any damage or accidents of which do not lead to significant damage to the reactor, but immediately lead to the cessation of plasma burning, i.e. to the reactor shutdown.

The refusal to use tritium in fuel (in the future) helps to reduce radiation hazard. The presence of tritium and the potential for its release into the environment do not allow us to confidently classify, for example, the ITER reactor as an internal safety facility. Internal safety is easily achieved by refusing to use tritium in fuel.

Despite the fact that the development of technology for thermonuclear installations with inertial plasma confinement lags noticeably behind reactors with magnetic confinement, significant steps have been made in the implementation of laser fusion reactor projects in recent years (Tables 10.1, 10.2, 10.3 and 10.4). Tables 10.1, 10.2 and 10.3 present the characteristics of spheromaks, large tokamaks and stellarators, respectively. Here R_1, R_2 are the major and minor radii of the plasma; B is the magnetic field strength on the plasma axis; I is the discharge current; τ is the pulse duration. The designation with an asterisk (*) corresponds to fuel is equal-component deuterium–tritium plasma, in other cases—deuterium plasma.

Since the mid-1980s, the world community has been working on the ITER (International Thermonuclear Experimental Reactor) project. ITER's goal is to demonstrate the possibility of commercial use of thermonuclear fusion reactions and to solve the physical and technological problems that may be encountered along the way. Construction is scheduled to be completed in 2035. ITER is being built in France (Saint-Paul-lès-Durance). It is planned to conduct experiments on the production of tritium from lithium-6 at the ITER tokamak reactor. ITER is focused on the use of D-T fuel [13]. The average thermonuclear power per pulse is 500 MW, the peak thermonuclear power in a pulse is 1100 MW. The pulse duration is about 400 s. The external plasma heating power is 73 MW. The time between pulses is about 1400 s [13].

The most advanced and powerful demonstration laser fusion facility (Table 10.2) NIF (National Ignition Facility) was built at the Lawrence Livermore National Laboratory (USA). A large research laser fusion facility, Laser Méga joule (LMJ), was

Table 10.1 Characteristics of spheromaks (spherical tokamaks)

Reactor (organization, country)	R_1/R_2, m	B,Tl	τ, s	I, MA	Year of launch (stop)
MAST (Nuclear Research Establishment, UK)	0.85/ 0.65	0.6	1–5	1.35 (2.0)	1999, 2020 (modernization)
NSTX (*Plasma Physics Laboratory, Princeton University, USA*)	0.85/ 0.65	0.3	1–5	1–2	1999 (on modernization)
Globus-M (Ioffe Physical-Technical Institute of the Russian Academy of Sciences, Russia)	0.36/ 0.24	0.65	< 0.3	0.35–0,5 (1.0)	1999

Table 10.2 Characteristics of large tokamaks

T-10 (National Research Center "Kurchatov Institute", Russia)	1.50/0.36	4.0	0.8	0.6	1978
T-15 (same place)	2.40/0.70	3.6	1.5	1.0	1988, 2023 (modernization)
PLT (*Plasma Physics Laboratory, Princeton University, USA*)	1.32/0.40	4.0	0.8	0.7 (1.0)	1975 (1996)
Doublet III (Institute for Plasma Physics, University of California, San Diego, USA)	1.45/0.45	2.6	2	2.2	1976 (1983)
Doublet III–D (same place)	1.66/0.67	2.2	5	3.0	1986
ET (University of California, Los Angeles, USA)	5.0/1.0	0.25 (1.0)	3	0.045	2000 (2006)
ASDEX (Max Planck Institute for Plasma Physics, Germany)	1.54/0.40	2.6	10	0.5	1980 (1990)
ASDEX Upgrade (same place)	1.65/0.5–0.8	3.9	5	1.4	1991
Textor (same place)	1.75/0.47	2.8	8–10	0.8	1991 (1994)
Tore Supra (Cadarache Research Center, France)	2.25/0.70	4.5	390	2.0	1998 (WEST—2016)
JT-60 (Institute for Nuclear Research, Japan)	3.00/0.70–1.00	4.5	7	2.1–3.2	1985 (2010)
JT-60U (same place)	3.40/1.16	4.2	28	5.0	1991 (1995)
HL-2A (Southwest Institute of Physics, China)	1.64/0.40	2.8	5	0.48	2002
EAST (Institute of Plasma Physics, Chinese Academy of Sciences, China)	1.75/0.43	5.0	1000	0.5	2006
KSTAR (National Institute for Fusion Research, Republic of Korea)	1.80/0.50	3.5	> 300	2.0 (4.0)	2008

(continued)

Table 10.2 (continued)

T-10 (National Research Center "Kurchatov Institute", Russia)	1.50/0.36	4.0	0.8	0.6	1978
TFTR (*Plasma Physics Laboratory, Princeton University, USA*)	2.1–3.1/0.4–0.96*	6.0	1,5	3.0	1982 (1997)
JET (Nuclear Research Establishment, UK)	2.96/0.96*	3.45–4.0	3–20	6.0	1983 (1992)
International project ITER (Cadarache Research Center, France)	6.2/2.0*	5.3	400	17.0	After 2035

Table 10.3 Characteristics of stellarators

LHD (National Institute of Fusion Science, Japan)	3.75/0*	3.0	3240 (120)	–	1998
Wendelstein 7-X (Max Planck Institute for Plasma Physics, Germany)	5.5/0.53*	3.0–6.0	1800	–	2016 (modernization)

Table 10.4 Laser thermonuclear fusion installations

Organization, country	Reactor	Year of launch	Number of bundles	Pulse duration, ns
Lawrence Livermore National Laboratory, USA	Nova	1984	10	1–3
	NIF	2010	192	20
University of Rochester, USA	Omega-2	1994	60	1–3
Naval Research Laboratory, USA	Nike	1994	56	4
Osaka University, Japan	GEKKO-XII	1981	12	0.5–2
All-Union Research Institute of Experimental Physics, USSR	Iskra-5	1989	12	0.25
Research Center "Limey", France	Phebus	1984	2	1–3

built by the French Nuclear Science Authority (Commissariat à l'Energie Atomique, CEA) near Bordeaux (France) [30]. The laser power is 1.8 MJ.

Neutron production at LMJ was demonstrated in 2019 [30]. A similar facility is being developed in Russia. NIAEP JSC is designing the UFL-2M laser thermonuclear fusion facility, which is planned to be built at RFNC-VNIIEF. The main objectives of such facilities are to develop and improve laser weapons.

NIF (National Ignition Facility) is the largest and most powerful facility in the world for experiments in the field of inertial thermonuclear fusion. In October 2010, the first and successful test of the world's largest NIF reactor for inertial controlled thermonuclear fusion took place. The NIF complex was built over 12 years. 192 laser beams are used to heat and compress a deuterium–tritium target with an outer radius of 1.8 mm, containing 150 μg of fuel. However, in 2012, after more than a thousand experiments, the NIF project was on the verge of closure. In 2018, after a series of technical improvements, the project managed to demonstrate a fusion reaction that released 3.6% of the laser input energy [31]. In December 2022, the NIF reactor achieved an excess of fusion power over the laser heating power of the target, i.e., in the mode of useful energy production [32].

In thermonuclear reactors (if tritium is not used as fuel), the ideals of internal safety are easily achieved. When tritium is used in thermonuclear fuel, the main danger in terms of tritium entering the environment may be the fuel production process.

Disruption of laser beam focusing will damage the installation (mainly electronic equipment), which will make the reactor impossible to operate.

Minimization of the activity of materials in thermonuclear reactors with magnetic or inertial plasma confinement is possible with the transition to neutron-free thermonuclear fusion (^{3}He–^{2}H).

Until 1991, all fusion reactors used deuterium fuel. An equal-component mixture of deuterium and tritium is the most promising fuel for energy production in future fusion power plants.

10.2.3 Self-sufficiency in Fuel

Deuterium reserves can be considered inexhaustible. The natural content of deuterium among stable hydrogen isotopes is 0.0115 ± 0.0070%. The world ocean contains a huge amount of deuterium—each cubic meter of sea water contains 33 g of deuterium. The volume of the world ocean is about 1340.74 million km^3, or about 1.34×10^{18} m^3. This means that the mass of deuterium in the world ocean is about 4.422×10^{16} kg.

When using the D-T fusion reaction, it is necessary to take into account that tritium is a short-lived nuclide: its half-life is 12.32 years. Tritium can be produced in a blanket of a thermonuclear reactor from lithium in a reaction under the action of neutrons generated by the fusion reaction:

$$^6\text{Li} + \text{n} \rightarrow \, ^3\text{H} + \, ^4\text{He}.$$

At relatively low neutron energies, the main mechanism of the $^{6}\mathrm{Li} + \mathrm{n}$ reaction is associated with the formation and decay of the compound nucleus $^{7}\mathrm{Li}^{*}$. (The asterisk denotes an excited state.) Eighteen decay channels of the compound nucleus $^{7}\mathrm{Li}^{*}$ have been well studied. The most probable of them are:

$$^{7}\mathrm{Li}^{*} \rightarrow {}^{7}\mathrm{Li} + \gamma;$$

$$^{7}\mathrm{Li}^{*} \rightarrow {}^{3}\mathrm{H} + {}^{4}\mathrm{He};$$

$$^{7}\mathrm{Li}^{*} \rightarrow {}^{6}\mathrm{He} + \mathrm{p}\ \left(\text{with subsequent } \beta \text{ - decay of } {}^{6}\mathrm{He} \text{ into } {}^{6}\mathrm{Li}\right);$$

$$^{7}\mathrm{Li}^{*} \rightarrow {}^{6}\mathrm{Li}^{*} + \mathrm{n}\ \left(\text{with subsequent decay of } {}^{6}\mathrm{Li} \text{ through different channels}\right),\ \text{etc.}$$

At the energy of incident neutrons below 20 MeV the decay channel of the compound nucleus $^{7}\mathrm{Li}^{*} \rightarrow {}^{3}\mathrm{H} + {}^{4}\mathrm{He}$ is not the main (most probable) one. The kinetic energy of neurons generated in the reaction $^{2}\mathrm{H} + {}^{3}\mathrm{H}$ is relatively small (about 14.1 MeV). According to various data the cross section of the reaction $^{6}\mathrm{Li}$ (n, $^{3}\mathrm{H}$) $^{4}\mathrm{He}$ is maximal (2.1 … 2.3 b) at the kinetic energy of the neutron about 5 MeV [19].

Thus, the possibilities for producing tritium are determined by the reserves of lithium in the earth's crust. Natural lithium consists of two stable isotopes: $^{6}\mathrm{Li}$ (6.5% by weight) and $^{7}\mathrm{Li}$ (93.5%). In some lithium samples, the isotopic ratio may be greatly disturbed due to natural or artificial fractionation of the isotopes. Separation of lithium isotopes may be necessary. Lithium-7 weakly absorbs neutrons. It can be used (possibly in alloys) as a coolant for fission reactors.

According to the US Geological Survey, global lithium resources identified as of 2023 total 98 million tonnes (up from 89 million tonnes in 2022), with reserves totaling 26 million tonnes.

So, if we compare the reserves of uranium and thorium on Earth taking into account the possibility of producing secondary fuel from $^{232}\mathrm{Th}$ and $^{238}\mathrm{U}$, then the total energy that can be produced by fission reactors and fusion reactors is of the same order.

10.2.4 Preferred Nuclear Reactions

The D-T reaction is characterized by a much larger cross section at relatively low energies of relative approach of the deuterium and tritium nuclei (Fig. 10.1). For the implementation of such a reaction, a temperature of ~ 10^{8} K and a Lawson number of 10^{14} cm^{-3} s are required.

The deuteron can interact with nuclei both by forming a compound nucleus and by direct interaction. This is due to the fact that the binding energy of the deuteron is about 2.2 MeV, i.e. about 1 MeV/nucleon, which is significantly less than the average specific binding energy (about 8 MeV/nucleon) of a nucleon in most nuclei. In addition, the average distance between a proton and a neutron in a deuteron is relatively large and is 4×10^{-13} cm. (For comparison: the average distance between nucleons in most nuclei does not exceed 2×10^{-13} cm).

If a deuteron forms a compound nucleus with a target nucleus, its excitation energy is approximately equal to 14 MeV due to the large difference in the values of the specific binding energy for the deuteron and most nuclei.

10.3 Results

10.3.1 Known Facts

The mechanism of fusion reactions is not important for the practical implementation of thermonuclear power plants, is of purely theoretical interest, and lies in the field of fundamental research. The mechanism of fusion reactions is usually not specified. It is known that two fundamentally different mechanisms of fusion reactions involving deuterium nuclei - deuterons are possible: the formation of a compound nucleus and incomplete penetration of a deuteron into a nucleus (the latter mechanism can formally be attributed to direct nuclear interaction). Generalized information on such reactions is presented in [23].

It is also known that during the formation of a compound nucleus, neutrons are more often present in the output channel of the D-D reaction, while protons are present in the case of direct interaction [23]. Moreover, in the reaction of incomplete penetration of a deuteron into a nucleus, the ratio of the number of protons and neutrons in the output channel of the D-D reaction decreases with an increase in the kinetic energy of the deuterons, while in the reaction with the formation of a compound nucleus, this ratio increases. This pattern is the main way to identify the mechanism of the D-D reaction.

10.3.2 New Mechanism of Synthesis of Deuterium and Tritium Nuclei

How to identify the mechanism of synthesis of deuterium and tritium nuclei at the kinetic energy of mutual approach characteristic of future power plants, if there are no protons in the reaction output channel? (Energy of $\approx$ 100 keV corresponds to the maximum reaction cross-section.) According to the data of work [23], D-T synthesis reactions are most often realized through the formation of a compound nucleus with high excitation energy, significantly exceeding the energy of separation of the nucleon and α-particle from the nucleus. But even in this case, there is reason to believe that the mechanism of the D-T reaction differs from the traditional mechanism of a compound nucleus proposed by N. Bohr [18].

What is the mechanism of the D-T reaction, which is always realized with the emission of neutrons and α-particles ($^2H + {}^3H \rightarrow {}^4He + n$)? According to the generalizations given in [9, 23], two options are possible: incomplete penetration of the deuteron into the tritium nucleus—triton (direct nuclear interaction) and the formation of a compound nucleus. A third option is also possible: a stripping reaction (of a proton of a deuterium nucleus), related to direct nuclear interaction. At the energy of mutual approach of D and T up to 200…300 keV, all these processes have the same output channel. The filling of the neutron shell in the magic tritium nucleus contributes to a decrease in the excitation energy of the compound nucleus 5He in the case of its formation, and of the products of direct nuclear interaction if the reaction occurs along this channel.

10.3.3 Facts in Favor of Direct Nuclear Interaction of Deuterium and Tritium Nuclei

We note the known experimental facts that formally do not allow us to classify D-T synthesis as a reaction realized through the formation of a compound nucleus (at energies of relative approach of deuterium and tritium nuclei of ~ 100 keV) [18, 23, 32]:

- The time of the D-T reaction only slightly (by an order of magnitude) exceeds the characteristic time τS of the strong (nuclear) interaction ($\tau_S \sim 10^{-23}\ldots10^{-22}$ s) and is several orders of magnitude shorter than the characteristic time τ of nuclear reactions realized through the formation of a compound nucleus (on average τ ~ 10–14…10^{-12} s).
- At high kinetic energy of relative approach of deuterium and tritium nuclei (~ 100 keV), the mean free path of, for example, a deuteron in nuclear matter (deuterium or tritium nucleus) significantly exceeds the characteristic size of another nucleus and the formation of a compound nucleus is theoretically impossible.
- The compound nucleus mechanism does not allow an adequate description of nuclear reactions at high excitation energy.
- The formation of a compound nucleus is usually supported by the resonant nature of the interaction cross-section dependence on the kinetic energy of the colliding particles. There are no clearly defined resonances in Fig. 10.1.
- The reaction has a single outlet channel.

The mechanism of direct interaction of deuterium and tritium nuclei is simple and clear, although the D-T fusion reaction does not quite fall under the definition of direct nuclear interactions. In direct interactions, the incident particle interacts only with a small number of nucleons of the nucleus. The energy introduced into the atomic nucleus is transferred predominantly to one or a small group of nucleons, the reaction occurs on the periphery of the nucleus. Even if the D-T fusion reaction is realized through the channel of direct nuclear interaction, then practically all nucleons of deuterium and tritium are involved in it, although two paired neutrons in the tritium nucleus do not participate in the reaction (redistribution of nucleons and energy between nucleons) and, being paired, pass into the ^{4}He nucleus. The protons in the deuterium and tritium nuclei pair up and form a stable compound in the new ^{4}He nucleus. Thus, the direct interaction reaction mainly concerns only two protons, initially located in different atomic nuclei. But this is only "mainly". The paired neutrons of tritium, one way or another, participate in the nuclear attraction of the proton in the deuterium nucleus, i.e., strictly speaking, all the tritium nucleons and the deuterium proton (half of the deuterium nucleons) participate in the interaction. Thus, the direct nuclear D-T interaction has its own specifics: the participation of almost all the nucleons in the colliding nuclei.

At relatively high energies E (several MeV, see Fig. 10.2), when the de Broglie wavelength λ of the deuteron or triton ($\lambda \sim h\ (2\ mE)^{-0.5}$; h is Planck's constant; m is the mass of the deuteron or triton) is less than the characteristic size of the nucleus, the micro cross section of the reaction of direct nuclear interaction of deuterium and tritium nuclei can be estimated as $\sigma \approx \pi\ (R_D{}^2 + R_T{}^2)$, where R_D and R_T are the radii of the deuterium and tritium nuclei, respectively [18]. If the characteristic size of the deuterium nuclei (1.963 fm, 1 fm $= 10^{-15}$ m) and tritium (less than 1.9 fm) is taken as the radius, then the maximum cross section $\sigma \approx 0.23$ b, which is noticeably less than the characteristic cross sections of the D-T reaction at energies of the order of 100 keV (see Fig. 10.1). It is known that the cross section of direct interaction is practically independent of the kinetic energy of the colliding particles.

Thus, there is reason to believe that direct nuclear interaction of deuterium and tritium nuclei dominates only at energies significantly exceeding (at least by one or two orders of magnitude) the characteristic values for fusion plants.

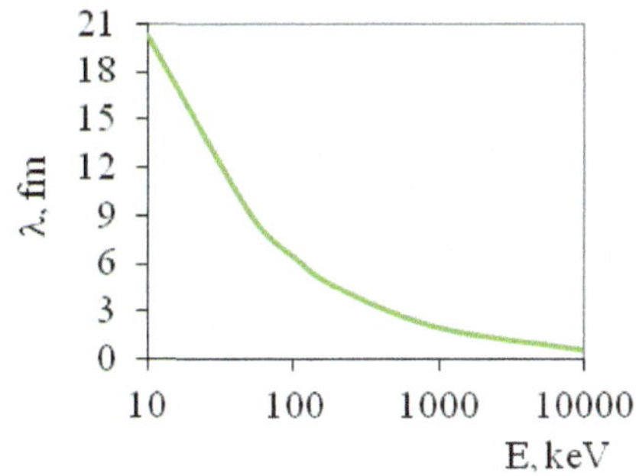

Fig. 10.2 Dependence of the de Broglie wavelength of the deuteron on its kinetic energy

10.3.4 Facts Confirming the Formation of a Compound Nucleus in D-T Fusion

According to the Bohr model, the nuclear reaction occurs in two stages. In the first stage, the initial particles form an intermediate (compound) nucleus during the time of the strong (nuclear—as a manifestation of the strong) interaction, i.e. the time required for the particle to cross the nucleus, approximately equal to $10^{-23} \ldots 10^{-21}$ s. In this case, the compound nucleus is always formed in an excited state, since it has excess energy brought by the particle into the nucleus in the form of the binding energy of the nucleon in the compound nucleus and part of its kinetic energy, which is equal to the sum of the kinetic energy of the target nucleus with mass number A and the particle in the center of mass system. The compound nucleus lives significantly longer than the characteristic time of the strong (nuclear—as a manifestation of the strong) interaction, and then, in the second stage, decays through some channel.

Let us note the experimental facts in favor of the formation of the compound nucleus ^{5}He at energy of mutual approach of the nuclei of deuterium and tritium of ~ 100 keV [18, 22, 23, 30, 32].

1. The compound nucleus really exists, but its average lifetime is extremely short.
2. The absence of clearly defined resonances may be due to overlapping levels in nuclei with high excitation energy, and in this case the dependence of the cross sections on the kinetic energy of the colliding nuclei has a fine structure, i.e. consists of many overlapping resonance bursts. The dependence for the D-T reaction shown in Fig. 10.1 is nothing more than a broad (giant) resonance with a fine structure. The cross section of the D-T reaction at energies of ~ 100 keV is more than an order of magnitude greater than the value corresponding to direct nuclear interaction and depends on the kinetic energy of the mutual approach of the nuclei.
3. According to the definition, a compound nucleus is a long-lived state that decays with a certain probability through different channels.

Let's consider these facts in more detail.

1. It is generally accepted that an atomic nucleus exists if the average time τ of its lifetime is not less than the characteristic time τ_S of the strong (nuclear) interaction, i.e. $\tau \geq \tau_S \sim 10^{-23} \ldots 10^{-22}$ s [9]. It is also generally accepted that a nucleus exists if it does not emit nucleons from the ground energy state, in other words, if the binding energy of the last neutron and/or proton is negative [9]. The characteristic lifetimes of compound nuclei are also known: $\tau \sim 10^{-14} \ldots 10^{-12}$ s. The decay of a compound nucleus is in many ways similar to the phenomenon of radioactivity—time-stabilized decay. However, unlike radioactivity, among the factors stabilizing the decay of a compound nucleus, the determining one is the effect described as one of the interpretations of the Heisenberg uncertainty principle, according to which the average lifetime τ of a compound nucleus and the width Γ of the energy level corresponding to a given excited state of the compound nucleus are related by the relation $\Gamma\tau \sim h$, where h is Planck's

constant. As a result, for example, the decay of a compound nucleus with the emission of a neutron (characterized by the neutron width Γ_n of the level) is realized not in a time of $\tau \sim \tau_S$, or slightly exceeding τ_S, but in a much longer time: $\tau_n \sim 10^{-18}...10^{-12}$ s [9]. The upper estimate corresponds to heavy nuclei, the lower one to light nuclei. The neutron width of the level for light nuclei is approximately the same: $\Gamma_n \sim 1$ keV, the decay time of a compound nucleus through the channel with the emission of a neutron is $\tau_n \sim h/\Gamma_n \sim 10^{-18}$ s.

It is known that the emission of neutrons by nuclei occurs due to strong (nuclear) interaction, and, despite some weakly influencing stabilization effects, quite quickly: in a time slightly exceeding τ_S, although there are known cases of neutron emission in $\tau_n \sim 10^{-17}$ or, at least, no more than 4×10^{-14} s [9].

In the fusion reaction of deuterium and tritium nuclei, the role of the compound nucleus is played by ^{5}He. This nucleus formally does not exist for the two above-mentioned reasons (i.e. by definition). Even in the ground energy state, the ^{5}He nucleus emits a neutron (whose binding energy is minus 0.96 MeV), and the half-life of the ^{5}He nucleus is $(7.00 \pm 0.30) \times 10^{-22}$ s [20], or $(6.02 \pm 0.22) \times 10^{-22}$ s [21], which corresponds to an average lifetime of $\sim 10^{-21}$ s, i.e., it is only an order of magnitude greater than τ_S. (The first successful attempts to prove the existence of ^{5}He were made in 1937, the first data on the nuclear-physical properties of ^{5}He were obtained by the 1950s and published in [33].)

The decay of $^5\text{He} \rightarrow {}^4\text{He} + \text{n}$ occurs due to nuclear interaction (the reaction products practically do not enter into electromagnetic interaction), i.e., in a time of $\tau \sim \tau_S$. This is the only decay channel of ^{5}He [20]. (A small stabilization is possible due to the emission of photons—quanta of electromagnetic interaction with a characteristic time of $\tau \sim \tau_{E\text{-}M} \sim 10^{-20} \gg \tau_S$.) Thus, it can be assumed that the D-T reaction can be realized through the formation of a compound nucleus of ^{5}He, but this nucleus quickly decays due to nuclear interaction (with a small stabilization in time due to electromagnetic interaction during the emission of γ-quanta):

$$^2\text{H} + {}^3\text{H} \rightarrow {}^5\text{He} \rightarrow {}^4\text{He} + \text{n} + 17.6\,\text{MeV},$$

i.e. formally (in accordance with the definition) the ^{5}He nucleus does not exist, but physically (in reality) it lives for a time τ_S, or slightly exceeding τ_S ($\sim 10^{-21}$ s). In any case, the ^{5}He nucleus is real, not virtual. As a result, the D-T synthesis process is much faster than most nuclear reactions that are realized through the formation of a compound nucleus. Moreover, the two stages of the D-T reaction: the formation and decay of the 5He compound nucleus cannot be considered independent as in the traditional model of N. Bohr's compound nucleus: they are realized in approximately the same time $\tau \sim \tau_S$ and the compound nucleus "remembers" the method of its formation.

Due to the absence of noticeable factors of stabilization of the decay of the compound nucleus ^{5}He, the reactions of D-T fusion, realized through the formation of a compound nucleus, and as direct nuclear interactions, can proceed in the same time. Moreover, D-T reactions realized through the formation of a compound nucleus can proceed faster than classical direct nuclear reactions with the emission of charged particles.

2. It is known that at high excitation energies exceeding the real (not average) binding energy of one or several nucleons in the nucleus, the energy levels overlap and there are no resonances in the energy dependence of the reaction cross section [32]. In these cases, it is possible to distinguish auto ionization levels corresponding to excitation energies exceeding the binding energy of a specific nucleon in the nucleus, preserving some properties of discrete energy levels. A compound nucleus in an excited state at one of the auto ionization levels decays due to nuclear interaction by emitting one or more nucleons. When protons are emitted, the decay of the compound nucleus is stabilized by electromagnetic interaction and proceeds more slowly than when neutrons are emitted. Auto ionization levels are usually observed at excitation energies of up to 30…50 MeV. They correspond to broad (so-called giant) resonances in the energy dependence of the cross sections [32].

The decay mechanism of the compound nucleus ^{5}He formed in the D-T fusion reaction is close to the model of decay from an auto ionization excited state. The fact that the ^{5}He nucleus emits a neutron even from the ground energy state allows us to extend the applicability of this mechanism to cases of relatively low excitation energies. This model of the behavior of the compound nucleus (decay from an auto ionization level) is realized in the absence of a Coulomb barrier.

To overcome the Coulomb repulsion of deuterium and tritium nuclei, the kinetic energy of their mutual approach must exceed 0.1 MeV [9]. (It should be noted that thermonuclear fusion reactions do not have a Coulomb barrier in the strict sense of this term [34].) As a result, in the presence of a Coulomb barrier in the D-T reaction, which proceeds with the formation of a compound nucleus, a scenario of decay of the compound nucleus ^{5}He is realized, close to the model of decay from an auto ionization excited state, and in the energy dependence of the reaction cross section there is a giant resonance (Fig. 10.1), which has a fine structure. Considering that the compound nucleus ^{5}He does not formally exist (it emits a neutron even from the ground state), this scenario turns out to be the only possible one.

3. Formally, it is possible to distinguish several output channels of the D-T fusion reaction, which is realized through the formation of a compound nucleus (see below).

10.4 Discussion. Output Channels of D-T Reactions

It is known that all processes in the physics of the micro world, including nuclear reactions in general, and the decay channels of a compound nucleus in particular, are probabilistic in nature. Then where does such certainty in determining the output channel of the D-T fusion reaction come from?

The ^{5}He nucleus formally does not exist for two reasons. First, although it is overloaded with only one neutron, this nucleus emits a neutron even from the ground energy state. Second, the average lifetime of the ^{5}He nucleus is comparable to the characteristic time of the strong interaction. If the reaction occurs through the formation of a compound nucleus ^{5}He, and other decay channels are prohibited by conservation laws, then the probability of ^{5}He decay through the channel with the emission of a neutron is 100% [21]. But there is no determinism here either, since different reaction output channels are possible, although a neutron and an α-particle are present in all of them with 100% probability: $^5\mathrm{He} \to {}^4\mathrm{He} + \mathrm{n}$; $^5\mathrm{He} \to {}^4\mathrm{He} + \mathrm{n} + \gamma$; $^5\mathrm{He} \to {}^4\mathrm{He} + \mathrm{n} + \gamma + \gamma$ etc. The decay of the compound nucleus ^{5}He via the inelastic scattering channel ($^5\mathrm{He} \to {}^2\mathrm{H} + {}^3\mathrm{H}$) is prohibited by the law of conservation of spatial parity in strong and electromagnetic interactions. In accordance with the Bohr mechanism, the decay of a compound nucleus via different channels is usually always accompanied by the emission of γ-quanta. The emission of γ-quanta in the D-T reaction is practically not observed. Gamma-quanta with energy from one to several MeV are produced, for example, in thermonuclear reactions:

$$\mathrm{p} + {}^3\mathrm{H} \to {}^4\mathrm{He} + \gamma;$$
$$\mathrm{p} + {}^2\mathrm{H} \to {}^3\mathrm{He} + \gamma;$$
$${}^3\mathrm{He} + {}^4\mathrm{He} \to {}^7\mathrm{Be} + \gamma.$$

There are also other reactions of γ-quanta production with energy of ≈1…5 MeV in nuclear fusion reactions. In 2020, scientists from JINR (Dubna, Russia) measured the yield of γ-quanta from nuclear fusion reactions coming from different spin states of the muon (μ) molecules pTμ and pDμ [35]. For the first time, fusion reaction channels with the yield of double γ-quanta were observed:

$$\mathrm{pT}\mu \to {}^4\mathrm{He}\mu + \gamma + \gamma;$$
$$\mathrm{pD}\mu \to {}^3\mathrm{He}\mu + \gamma + \gamma.$$

The yield of single γ-quanta is observed in the fusion reaction channel

$$pT\mu \to {}^4He\mu + \gamma[35].$$

If the reaction proceeds as a direct nuclear interaction, then the presence of a neutron and an α-particle in the output channel is determined only by the maximum probability at a kinetic energy of mutual approach of the D and T nuclei of up to 200…300 keV. At significantly higher energies, other channels of direct interaction open. At energies up to 200…300 keV, the cross sections of the D-T fusion reaction, proceeding through other channels, are so small that they are inaccessible to experimental observation [18].

Thus, in a collision of deuterium and tritium nuclei the following are possible: elastic scattering, which occurs at practically any kinetic energy of the colliding nuclei (not of interest for obtaining energy); direct nuclear interactions; the formation of a compound nucleus ^{5}He with its decay via the channel ^{4}He + n (at a kinetic energy of mutual approach of the nuclei D and T up to 200 keV).

At energies of mutual approach of deuterium and tritium nuclei of ≈ 200...300 keV, direct interactions of D and T nuclei are realized as incomplete penetration of the deuteron into the tritium nucleus or as a stripping reaction. In both cases, the output channel is ^{4}He + n. At energies significantly exceeding 300...400 keV, the probability of formation of a compound nucleus is negligibly small, direct interactions dominate, and other channels of direct interaction open up. The latter processes are in many ways analogous to reactions of deep splitting of atomic nuclei, although, obviously, they do not have the avalanche-like character inherent in spallation reactions [9].

The loosely bound deuterium nucleus (the total binding energy of the deuteron is 2.225 MeV) easily decays into nucleons in direct nuclear interaction.

Unlike the deuteron, the tritium nucleus is unstable (half life is 12.33 years ± 7.30 days [22]) despite the full neutron shell. The magic number of neutrons in tritium is an important stabilizing factor: tritium does not emit a neutron from the ground state (unlike ^{5}He). The instability of tritium is due to the overload of the nucleus with neutrons: one neutron is "extra". In direct reactions, the tritium nucleus also easily decays into nucleons if the interaction occurs with a neutron that is part of the deuteron, and the deuteron proton is at a relatively large distance (more than r_S, or at least greater than the distance between nucleons in the triton) from the tritium nucleus.

If the specific binding energy of two paired neutrons in a triton is relatively high (this is the reason why tritium does not emit a neutron from the ground state), then the binding energy of the proton in the triton is much lower, and the proton is easily knocked out of the nucleus by direct nuclear interaction with a neutron in deuterium (or with a proton at much higher energies). In the absence of a proton, the paired neutrons decay into two unbound neutrons.

If a triton enters into direct nuclear interaction with a deuterium proton, then even at high kinetic energy (200...300 keV) of the relative approach of the D and T nuclei, the probability of the formation of a doubly magic nucleus ^{4}He in the reaction output channel is high. With a further increase in energy, nucleons unbound into nuclei and light deuterium nuclei may dominate in the output channel. However, such processes are of no practical interest for obtaining energy, since the energy release is maximal if doubly magic nuclei are present in the reaction output channel, in this case, ^{4}He.

10.5 Conclusion

Based on the analysis of the results of experimental data on fusion reactions and the properties of ^{5}He nuclei obtained by different authors (some of them are combined into databases [20–22]), it can be concluded that the mechanism for the implementation of the D-T fusion reaction with the formation of a compound nucleus (at an energy of mutual approach of the deuteron and triton of ~ 100 keV) can formally be considered "intermediate" between direct nuclear interaction and the mechanism for the formation of a compound nucleus proposed by N. Bohr. ("Formally", since the ^{5}He nucleus also "formally" does not exist, although it does exist in reality for a time of ~ 10^{-21} s [20].) In any case, this is not a direct interaction. The reaction is realized through the formation of a compound nucleus, but is not described by the mechanism proposed by N. Bohr. In this case, the reaction time is characteristic of direct nuclear interactions, and the sequence of events is: the formation of a real, not virtual, compound nucleus, although formally non-existent, which quickly decays in a time slightly exceeding τ_S into ^{4}He and a neutron. The mechanism and energy dependence of the cross section are characteristic of nuclear reactions realized through the formation of a compound nucleus, with its decay from an auto ionization excited state.

Direct nuclear interactions of the deuteron and triton also differ somewhat from the classical ones, since all nucleons of the colliding nuclei participate in one way or another in the D-T interaction.

The proposed new mechanism of the D-T fusion reaction is one of the main current issues in plasma physics and thermonuclear fusion.

References

1. T. Shelby, *The First Reactor* (U.S. Department of Energy, DOE/NE-0046, Springfield, Virginia, 1982). https://www.energy.gov/sites/prod/files/The%20First%20Reactor.pdf. Accessed 3 Jan 2025
2. E. Fermi, Experimental production of a divergent chain reaction. Am. J. Phys. **20**, 536 (1952). https://doi.org/10.1119/1.1933322
3. C. Chunsheng, Environmental problems of the nuclear industry and the situation in China. AtomInfo.Ru, 1 Nov (2007). http://atominfo.ru/news/air2484.htm. Accessed 3 Jan 2025
4. The Database on Nuclear Power Reactors. IAEA Power Reactor Information System. http://www.iaea.org/pris. Accessed 3 Jan 2025
5. V.L. Ginzburg, On superconductivity and superfluidity (what I managed to do and what I failed to do), as well as on the "physical minimum" at the beginning of the 21st century. Phys. Usp. **47**(11), 1155–1170 (2004). https://doi.org/10.1070/PU2004v047n11ABEH001825
6. V.L. Ginzburg, What problems of physics and astrophysics seem especially important and interesting now? Soviet Physics Uspekhi **14**(1), 21 (1971). https://doi.org/10.1070/PU1971v014n01ABEH004434
7. Sir Rutherford, E.F.R.S.: Collision of α particles with light atoms. An anomalous effect in nitrogen. The London, Edinburgh, and Dublin Philosophical Magazine and Journal of Science, Series 6, 37(222), 581–587 (1919), https://doi.org/10.1080/14786440608635919
8. N. Bohr, Neutron capture and nuclear structure. Usp. Fiz. Nauk **16**(4), 425–435 (1936). https://doi.org/10.3367/UFNr.0016.193604a.0425
9. Okunev, V.S.: Fundamentals of Applied Nuclear Physics and Introduction to Nuclear Reactor Physics. Textbook. Series: Physics at the Technical University. 2nd edn. Moscow, Publishing House of Bauman Moscow State Technical University (2015).
10. 22nd IAEA Fusion Energy Conference celebrating fifty years of fusion entering into the burning plasma era, 13–18 Oct 2008, Geneva, Switzerland (2008). http://www-pub.iaea.org/mtcd/meetings/Announcements.asp?ConfID-165. Accessed 3 Jan 2025

11. 14th International Conference on Fusion Reactor Materials ICFRM-14, 10–11 Sept 2009. Sapporo, Japan (2009), http://www-pub.iaea.org/mtcd/meetings/Announcements.asp?ConfID=39073. Accessed 3 Jan 2025
12. 23rd IAEA Fusion Energy Conference. 11–16 Oct 2010. Daejon, Republic of Korea (2019), http://www-pub.iaea.org/mtcd/meetings/Announcements.asp?ConfID=38091. Accessed 3 Jan 2025
13. ITER Physics Basis. Nuclear Fusion 39 (12) (1999), https://iopscience.iop.org/issue/0029-5515/39/12, https://doi.org/10.1088/0029-5515/39/12/301, https://doi.org/10.1088/0029-5515/39/12/309. Accessed 3 Jan 2025
14. K. Ikeda, Progress in the ITER physics basis. Nucl. Fusion **47**(6), E01 (2007). https://doi.org/10.1088/0029-5515/47/6/E01
15. M. Keilhacker, J.M. Adams, H. Altmann, A. Andersen, P. Andrew, M. Angelone, W. Zwingmann, JET: progress in performance and understanding. Plasma Phys. Controlled Fusion **33**(13), 1453–1478 (1991)
16. G. Taylor, J.R. Wilson, R.C. Goldfinger, J.C. Hosea, D.J. Hoffman, R. Majeski, S. Zweben, ICRF Heating of TFTR Deuterium Supershot Plasmas in the ^{3}He Minority Regime. Plasma Phys. Controlled Fusion **36**(3), 523–542 (1994). https://doi.org/10.1088/0741-3335/36/3/010
17. Kikuchi, M., Lackner, K., Tran, M.Q. (eds.): Fusion physics. International Atomic Energy Agency, Vienna, Austria (2012), https://www-pub.iaea.org/MTCD/Publications/PDF/Pub1562_web.pdf. Accessed 2025/01/03.
18. Okunev, V.S.: On the mechanism of the nuclear reaction D-T fusions. Herald of the Bauman Moscow State Technical University. Series Natural Sciences 2, 18–28 (2012).
19. Pronyaev, V.G.: Lithium-6. Encyclopedia of neutron data ROSFOND (Russian library of files of evaluated neutron data). Obninsk, IPPE (2006), https://z-lib.io/book/16496837; https://www.ippe.ru/reactors/reactor-constants-datacenter/rosfond-neutron-database. Accessed 2025/01/03.
20. G. Audi, F.G. Kondev, M. Wang, W.J. Huang, S. Naimi, The NUBASE2016 evaluation of nuclear properties. Chin. Phys. C **41**(3), 030001 (2017). https://doi.org/10.1088/1674-1137/41/3/030001
21. F.G. Kondev, M. Wang, W.J. Huang, S. Naimi, G. Audi, The NUBASE2020 evaluation of nuclear physics properties. Chin. Phys. C **45**(3), 030001 (2021). https://doi.org/10.1088/1674-1137/abddae
22. Chadwick, M.B., Herman, M., Obložinský, P., Dunn, M.E., Danon, Y., Kahler, A.C., Young, P.G.: ENDF/B-VII.1 Nuclear Data for Science and Technology: Cross Sections, Covariances, Fission Product Yields and Decay Data. Nuclear Data Sheets 112 (12), 2887–2996 (2011), https://doi.org/10.1016/j.nds.2011.11.002.
23. Mukhin, K.N.: Experimental Nuclear Physics: Textbook for Universities. In 2 books. Book 1. Physics of the Atomic Nucleus. Part II. Nuclear Interactions. Moscow: Energoatomizdat (1993).
24. Thomson, G.P., Blackman, M.: Patent Specification GB817681: Improvements in or relating to gas discharge apparatus for producing thermonuclear reactions. European Patent Office (August 6, 1959).
25. Bondarenko, B.D.: Role played by O.A. Lavrent'ev in the formulation of the problem and the initiation of research into controlled nuclear fusion in the USSR. Physics-Uspekhi 8, 886–894 (2001), https://doi.org/10.1070/pu2001v044n08abeh000910.
26. Tanner, E.C.: Project Matterhorn: An Informal History. Princeton University Plasma Physics Laboratory (1982).
27. Kletz, T.: What You Don't Have, Can't Leak, Chemistry and Industry, 287–292 (1978).
28. Kletz, T.A.: Cheaper, Safer Plants or Wealth and Safety at Work – Notes on Inherently Safer and Simpler Plants. Rugby, Institution of Chemical Engineers (1984).
29. I. Spiewak, A.M. Weinberg, Inherently Safe Reactors. Annu. Rev. Environ. Resour. **10**, 431–462 (1985). https://doi.org/10.1146/annurev.eg.10.110185.002243
30. LMJ-PETAL User Group. LMJ-PETAL Academic Access Experiments. CEA DAM (2024). https://www-lmj.cea.fr/LMJ-PETAL-User-Group.html. Accessed 3 Jan 2025

31. J. Crippen, D. Casey, D. Edgell, D.N. Fittinghoff, M. Gatu-Johnson, C. Goyon, S. Haan, O.A. Hurricane, Fusion energy output greater than the kinetic energy of an imploding shell at the national ignition facility. Phys. Rev. Lett. **120**, 245003 (2018). https://doi.org/10.1103/PhysRevLett.120.245003
32. DOE National Laboratory Makes History by Achieving Fusion Ignition, Energy.gov, 13 Dec 2022. https://www.energy.gov/articles/doe-national-laboratory-makes-history-achieving-fusion-ignition. Accessed 3 Jan 2025
33. B.R. Mottelson, Elementary types of excitation in nuclei. Soviet Phys. Uspekhi **120**(12), 563–580 (1976). https://doi.org/10.3367/UFNr.0120.197612c.0563
34. K. Artemov, B. Dzhelepov, Radioactive isotopes of helium. Soviet Physics Uspekhi **61**(2), 189–210 (1950). https://ufn.ru/ufn50/ufn50_6/Russian/r506_d.pdf. Accessed 3 Jan 2025
35. S.Y. Lukyanov, N.G. Kovalsky, *Hot Plasma and Controlled Thermonuclear Fusion* (MEPhI, Moscow, 1997)
36. A. Adamchak, V.A. Baranov, L.N. Bogdanova, V.P. Volnykh, O.P. Vikhlyantsev, S.S. Gershtein, A.A. Yukhimchuk, Yield of γ-quanta from the nuclear fusion reaction in the muonic molecules ptμ and pdμ. JETP **161**(2), 177–183 (2022). https://doi.org/10.31857/S0044451022020031

Chapter 11
Research of the Physical Basis of Heat Transfer in Heat Exchanger Devices

Saydulla Khuzhakulov, Zakir Pardaev, Ismoildjon Muradov, and Timur Shodiev

Abstract The purpose of this article is to determine the dependence of heat transfer processes on balance and criterion equations, heat transfer in external and internal heat exchange, criterion equations of heat transfer in heat exchange devices, the dependence of heat on physical parameters, and mathematical modeling of the process. The heat transfer coefficient was calculated based on criterion equations related to the Nusselt, Prandtl (from the table), Grasshoff, Pecle and Reynolds numbers. When a pipe with d_0 = 32 mm was studied, according to the calculations based on criteria, it was determined that the heat transfer coefficient is 14.5×10^3 W/m^2 K and the heat flux is 29.14×10^3 W when the Nusselt number is 730.47. When the heat exchange is calculated when the flow crosses a large number of pipe bundles, the Reynolds number for the flow at a speed of 4 m/s is equal to 12.7×10^4, that is, the flow is turbulent. According to the criterion equations, the Nusselt number reaches 749.7, and the value of the heat transfer coefficient reaches 14,021 W/m^2 K. Based on the results of mathematical modeling based on the selected boundary conditions, it was determined that the coefficient of heat transfer is greater, it depends on the flow regime of heat carriers, Reynolds, Nusselt criteria and Prandtl number, and graphs of dependence were obtained.

11.1 Introduction

Heat exchangers are an integral part of industrial processes, especially heating, ventilation, and cooling processes in thermal power plants, due to their high efficiency in transferring heat between working bodies [1–4]. In terms of their structural design, concentric tube heat exchangers are distinguished by their simplicity, economy, and widespread use [5, 6]. The performance of heat exchangers is significantly affected by their geometric parameters and flow conditions [7, 8]. Factors such as the inner diameter of the pipe, the ratio of diameters, the length, and the Reynolds number

S. Khuzhakulov (✉) · Z. Pardaev · I. Muradov · T. Shodiev
Karshi State Technical University, Karshi, Uzbekistan
e-mail: bobur160189@mail.ru

© The Author(s), under exclusive license to Springer Nature Switzerland AG 2025

A. Gibadullin et al. (eds.), *Proceedings of the 4th International Scientific and Practical Symposium on Materials Science and Physics*, Springer Proceedings in Physics 318,
https://doi.org/10.1007/978-3-031-94805-3_11

of the hot and cold coolants can significantly affect the heat transfer coefficient and pressure differences in the heat exchanger.

Furthermore, the correct choice of the working medium has a significant impact on the thermal characteristics [9, 10], as the selection of different heat carriers (e.g., air–air, air–water, water–air, and water–water) can change the thermal and hydraulic characteristics of the system and thus affect the overall efficiency of the system.

11.1.1 Statement of the Problem

In contrast to the concept of heat conduction, when heat energy is transferred from a stationary or moving medium through its molecular structure, the movement of the medium causes heat transfer. A moving medium is understood to be a liquid or heat carrier [11].

Thus, the convective transfer of heat between layers of a heat carrier flow with different temperatures is called convective heat transfer. In practice, heat exchange between a heat carrier flow and a wall surface, i.e. convective heat transfer, is often observed. In this process, heat is transferred in one direction, i.e. either from the heat carrier to the wall surface, or vice versa, from the wall surface to the heat carrier flow. The purpose of this article is to determine the dependence of heat transfer processes on balance and criterion equations, heat transfer in external and internal heat exchange, criterion equations of heat transfer in heat exchange devices, determination of the dependence of heat on physical parameters, and mathematical modeling of the process.

11.2 Materials and Methods

The basic equation representing heat transfer is Newton's equation:

$$Q = \alpha \cdot \left(T_j - T_w\right) \cdot F_w, \quad q = \alpha \cdot \left(T_j - T_w\right), \tag{11.1}$$

here Q—heat flow in heat supply, W; T_j—average temperature of the heat carrier over the longitudinal heat exchange surface F_w; T_w—average temperature of the heat exchange surface; $T_j - T_w = \Delta T_w$—temperature difference in heating; $q = Q/F_w$—heat flow density on the heat exchange surface, W/m^2; α—heat transfer coefficient, W/(m^2 K).

Heat transfer problems are solved by calculating the given heat flux Q at given values of the heat exchange surface F_w, or, conversely, by determining the required surface F_w at which heat is transferred at a given value of the heat flux Q.

In general, the heat transfer coefficient is calculated based on the following equations:

$$Nu = f(Re, Gr, Pr, \ldots) \tag{11.2}$$

Derivation of criteria of constituent similarity in this expression is presented in educational and scientific literature [11–13].

From the expression (11.2), the Nusselt number is the quantity determined by calculation, which contains the required coefficient α:

$$Nu = \frac{\alpha \cdot l_0}{\lambda_j}, \tag{11.3}$$

here l_0—descriptive dimension of the process, $\lambda_j(T_j)$—is the thermal conductivity of the heat carrier and is determined as a physical parameter from references [16] and depending on its type and temperature from [10].

The next constituent, the numerical value of which is the Reynolds number, describes the flow regime of a heat carrier moving with a velocity w_j relative to the heat exchange surface:

$$Re = \frac{\rho_j \cdot w_j \cdot l_0}{\mu_j}, \tag{11.4}$$

The criterion equation describing the intensity of heat exchange in free convection conditions of the heat carrier is the Grasshof criterion:

$$Gr = \frac{\beta_j \cdot g_0 \cdot \Delta T_w \cdot l_0^3}{v_j^2}; \tag{11.5}$$

this expression takes into account the temperature difference between the coolant and the surface of the object ΔT_w and the acceleration of the environment; $v_j(T_j) = \frac{\mu_j}{\rho_j}$—kinematic viscosity of heat carriers; $\beta_j(T_j)$—volume expansion coefficient, for gases $\beta_\Gamma = \frac{1}{(T_j)}$.

This value is also taken from the reference books for liquids or calculated from the density chosen for a given temperature interval:

$$\beta_j(\Delta T_w) = \frac{(\rho_w - \rho_j)}{\rho_j}. \tag{11.6}$$

Along with the coefficients λ_j (T_j) and μ_j (T_j) of liquid heat carriers, depending on the nature and temperature of the heat carrier, the physical parameter Prandtl criterion Pr_j (T_j) is determined from the references (most often [13, 15, 16] is used for liquids and gases).

To perform heat calculations, the heat physical properties of heat carriers can be determined from analytical relationships over a specified temperature range [15].

Often, complex similarity numbers are used in criterion equations, such as the Peclet number $Pe = Re \cdot Pr$, the Reley number $Ra = Gr \cdot Pr$.

The shape factor of the body Φ—takes into account all the features of the heat carrier flow around it, depending on the size and shape of the surface of the body.

When developing and calculating a mathematical model of the heat transfer process, the recommended governing equations of heat transfer help to generalize the experimental results by a maximum of 10–20%. When solving a specific problem, when selecting governing equations from reference books, it is necessary to pay attention to the following:

- Depending on the external factors that cause the movement of the heat carrier, it is divided into free and forced convection. However, there may also be a mixed convection case;
- Forced convection can be performed for cases of unrestricted flow of the heat carrier and movement of the heat carrier in pipes and channels. Calculation of unrestricted flow is an external problem, calculation of the heat carrier moving through the pipe is an internal problem [17];
- Depending on the intensity of movement, forced convection can occur in laminar or turbulent modes of the heat carrier.

The governing equations of heat transfer, depending on the given and a number of other features, differ fundamentally from each other. Accordingly, the choice of a governing equation for a specific case of heat transfer is of great importance in solving the problem posed. The similarity criteria included in Eq. (11.2) are the physical parameters of heat transfer of liquid and gaseous heat carriers. The method of accurately estimating the parameters in the process is very complicated. In technical calculations, the method of presenting experimental data and calculation results in tabular form is widespread. It should be taken into account that the data given in different reference books for the same type of working body may differ in value. To avoid this situation, it is advisable to refer to the reference materials used in the calculation practice.

When selecting the physical parameters of heat carriers from reference tables, it is necessary to distinguish between the given values from a certain range of values. In this case, for accurate calculations, it is necessary to use the interpolation method. For example, to determine the parameter A based on the given parameter B, the following formula is used:

$$A = A_q + \left(A_{up} - A_l\right) \cdot \frac{B - B_l}{B_{up} - B_l}, \tag{11.7}$$

where A is the parameter $A_{up} < A < A_l$ value to be searched between the upper and lower (larger and smaller) values in the table; B is a given parameter of the function A(B) and its value is found in the interval $B_{up} < B < B_l$.

In external heat transfer problems, the criterion equation in the form of Eq. (11.2) for calculating heat transfer by free convection over a body in a uniform unconfined isobaric environment does not take into account the criterion number Re for the case where there is no forced motion of the heat carrier, and the dynamics of the motion can be fully described by the magnitude of the criterion Gr. It is customary to call the product of the criteria Gr · Pr = Ra the Reley number, in which case expression (11.2) is written as follows [11, 13]:

$$Nu_p = C \cdot (Ra_p)^n \cdot \left(\frac{Pr_p}{Pr_w}\right)^{0,25}, \tag{11.8}$$

where C is a constant and the power index n is a function of the number Ra and is determined from the numerical values and depending on the shape of the heat exchange surface as follows:

- For vertical walls $Ra_p = 1 \times 10^3$ to 1×10^9, C = 0.75–0.8; n = 0.25;
- For vertical walls $Ra_p > 6 \times 10^9$, C = 0.15; n = 0.333;
- For horizontal pipes $Ra_p = 1 \times 10^3$ to 1×10^9, C = 0.50; n = 0.25.

The index "p" in the expression denotes the physical properties of the heat carrier (β, λ, ν, Pr) related to the similarity criterion. These values are chosen as the average value based on the calculated temperature of the moving heat carrier $t_p = ((t_w + t_j))/2$.

The following expression is used to calculate the average heat transfer coefficient of the vertical surface under the conditions of q_w = const:

$$Nu_j = 0.75 \cdot Ra_j^{0,25} \cdot \left(\frac{Pr_j}{Pr_w}\right)^{0,25}, \tag{11.9}$$

in this case, the temperature of the heat carrier at the boundary of the moving layer is taken. When determining the heat transfer over a horizontal cylindrical pipe, the constant value in the expression is taken to be equal to 0.5.

Equation (11.8) includes a correction for the number Pr during temperature changes and, at the same time, for the physical parameters of the thickness of the moving layer formed on the heat exchange surface [11, 12].

Heat transfer under forced convection conditions. The problems of heat transfer in this case are also worked out on the basis of the criterion Eq. (11.2). The structure and constants of this equation are determined from the flow regime of the heat carrier relative to the heat exchange surface. Also, based on the boundary conditions of the process, the problem of heat transfer is divided into external and internal problems [11, 13].

Internal heat transfer problems are characterized by the flow of the heat carrier inside pipes and channels, its interaction with the wall surface and filling the horizontal surface of the channel. In this case, the flow regime is divided into laminar $Re \leq 2300$, turbulent $Re > 10^4$ and transitional $2300 \leq Re < 10^4$ regimes.

Also, the laminar flow regime is divided into viscous ($Re \leq 2300$, $Ra \leq 8 \times 10^5$), in which natural convection does not significantly affect heat transfer, and viscous-gravitational ($Re \leq 2300$, $Ra > 10^6$), in which free convection affects the heat carrier layers. In the process of analyzing scientific works and studying scientific and technical literature, work was carried out to apply Eq. (11.2) in real situations and calculate it.

In deaerators, pyrolysis plants, device for recycling municipal solid waste [14] and condensers, the process of heat transfer of the coolant in pipes and channels under forced convection conditions is considered.

The most important heat transfer process in all thermodynamic processes is the heat transfer of the working bodies moving in the pipes.

The calculation of the heat transfer of the coolant flow in filled channels or pipes is an internal issue of the heat transfer process, and in such cases the following characteristic parameters of the process are selected [12, 13]:

- Equivalent or hydraulic diameter of the channel d_e:

$$d_e = \frac{4F_0}{P}, \tag{11.10}$$

where F_0 is the cross-sectional area of the channel (or the cross-sectional area through which the heat carrier flows); P is the perimeter of the cross-sectional area through which the heat carrier flows;

- The flow velocity in the channel w_m (or mass velocity), which is found from the flow rate of the heat carrier:

$$G = \rho \cdot w_m \cdot F_0, \quad \text{where } w_m = \frac{G}{(\rho \cdot F_0)}, \frac{\text{m}}{\text{s}}. \tag{11.11}$$

The universal formula for calculating the Reynolds number for internal flows, taking into account expression (11.11), is written as follows:

$$Re = \frac{\rho \cdot w_m \cdot d_e}{\mu} = \frac{4G}{\mu \cdot P}, \tag{11.12}$$

where μ is the dynamic viscosity of the heat carrier, its value is selected from the reference according to the temperature value T [15, 16].

The choice of the heat transfer criterion equation of the form (11.2) and its appearance as a function of the number Re is shown in the calculation of complex heat processes.

11.2.1 Recommendations

Recommendation 1. When calculating the heat transfer of heat-carrying fluids, special attention should be paid to the Prandtl number $Pr(T_j)$. Its value, unlike gases, is equal to 100 and sometimes 1000, and varies depending on the temperature of the heat-carrying fluids. The change in the Pr number is taken into account by the correction $(Pr_j/Pr_w)^{0.25}$.

Recommendation 2. To determine the heat transfer in internal heat exchange problems, a change is made to the hydrodynamic stability section ε_l. This value takes into account the influence of the heat carrier entering the pipe on the intensive heat exchange. The correction value depends on the flow regime of the coolant and the relative size of the pipe (l/d_0) and is given in the form of a table and formula in the reference books [15].

Recommendation 3. The choice of the criterion equation for heat transfer depends on the exact calculation conditions.

11.3 Results and Discussions

When water is pumped at a speed of $w_j = 4$ m/s in a 1-m-long HE pipe channel, the effect of the shape of the channel's cross-sectional surface on the heat transfer coefficient and the magnitude of the heat flux transferred from the wall to the water flow is studied and modeled for

- A pipe with a diameter of $d_0 = 32$ mm;
- A rectangular channel with an aspect ratio of 1:25.

The average water temperature along the length of the channel is $t_m = 40$ °C and the channel wall temperature $t_w = 80$ °C, and the channels are assumed to have the same cross-sectional area F_0 and water flow velocity w_j (Fig. 11.1).

Based on the average water temperature $t_m = 40$ °C, we obtain the physical parameters of heat from [4]:

$$\nu = 0.658 \times 10^{-6}\,\frac{\text{m}^2}{\text{s}};\quad \lambda = 0.635\,\frac{\text{W}}{\text{m K}};\quad Pr_j = 4.31;$$

$$t_w = 60\,°\text{C} - Pr_w = 2.99.$$

For case 1: we determine the equivalent diameter of the channel:

$$d_э = \frac{4 \cdot F}{P} = \frac{4}{\pi} \cdot \frac{\pi}{4} \cdot \frac{d_0^2}{d_0} = d_0 = 32\,\text{mm},$$

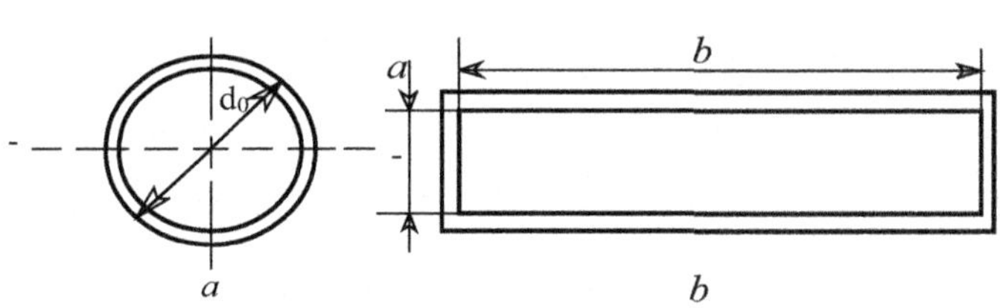

Fig. 11.1 1-m-long pipe and rectangular channel heat exchanger

- The equivalent diameter of a circular pipe is equal to its internal diameter.

To determine the flow regime of water in the pipe, we calculate the criterion equation Re:

- The equivalent diameter of a circular pipe is equal to its inner diameter.

To determine the flow regime of water in the pipe, we calculate the Re criterion equation:

$$Re = \frac{w_j \cdot d_0}{\nu_j} = \frac{4 \cdot 0.032}{0.658} \cdot 10^6 = 19.45 \cdot 10^4 > 10^4$$

i.e. the flow regime is turbulent.

Reference [15] the criterion equation for the heat transfer coefficient is selected from the literature and calculated. For the case where Pr > 0.7 in the turbulent movement of heat carriers along the pipe, it is recommended to use the following criterion equation:

$$Nu_j = 0.021 \cdot Re_j^{0.8} \cdot Pr_j^{0.43} \cdot \psi(Pr) \cdot \varepsilon_l, \tag{11.13}$$

where $\psi(Pr) = \left(\frac{Pr_j}{Pr_w}\right)^{0.25} = \left(\frac{4.31}{2.99}\right)^{0.25} = 1.09$ is the correction factor that determines the change in the temperature of the heat carrier on the cross-sectional surface of the pipe, since the Pr number is highly dependent on temperature for all flowing fluids, and in addition, the temperature range of heat carriers is also large.

The correction factor for the initial flow point of the heat carrier in the pipe also has the value $\varepsilon_l < 1$, subject to the condition $\frac{l}{d_0} = \frac{1.0}{0.032} = 31.25 < 50$ [14].

$$Nu_j = 0.021 \cdot \left(19.45 \cdot 10^4\right)^{0.8} \cdot 4.31^{0.43} \cdot 1.09 = 730.47;$$
$$\alpha = \frac{Nu_j \cdot \lambda}{d_0} = \frac{730.47 \cdot 0.635}{0.032} = 14.5 \times 10^3 \frac{\text{W}}{(\text{m}^2\,\text{K})}.$$

We determine the heat flow during heat transfer:

$$Q_0 = \alpha \cdot F_w \cdot (t_w - t_m) = 14.5 \cdot 10^3 \cdot 3.14 \cdot 0.032 \cdot 1.0 \cdot (60 - 40)$$
$$= 29.14 \cdot 10^3\ \text{W}.$$

For case 2: since the dimensions of the channel passage surface are not given, further studies are carried out using the relative method. For channels with the same cross-sectional area, the following can be written:

$$F_0 = a \cdot b = \frac{\pi \cdot d_0^2}{4} = 25 \cdot a^2, \tag{11.14}$$

because according to the initial conditions a:b = 1:25. Finally $a = \left(\frac{\pi}{100}\right)^{0.5} \cdot d_0 = \left(\frac{1}{100 \cdot \pi}\right)^{0.5} \cdot \pi d_0$.

To calculate the perimeter P of the channel, we find the following relationship:

$$P_0 = \pi \cdot d_0; \;\; P_k = 2 \cdot (a + b) = 2 \cdot a \cdot \left(1 + \frac{b}{a}\right) = 52a.$$

$\frac{P_k}{P_0} = 52 \cdot \left(\frac{1}{100 \cdot \pi}\right)^{0.5} = 2.93$, i.e., in a channel with the same cross-sectional area, the perimeter of the channel with an aspect ratio (a:b = 1:25) is almost three times larger than the perimeter of the pipe section.

To compare the intensity of heat transfer, the ratio of the Re numbers in the criterion equation is found. The answer to the question of whether the water flow regime in the channels changes or not is found.

We determine the Re number through the coolant flow:

$$Re = \frac{4 \cdot G}{\mu \cdot \Pi}, \tag{11.15}$$

where $G = \rho \cdot w_j \cdot F_0 = const$, because according to the initial conditions, $w_j = const$, $F_0 = const$, and also if $\rho = const$, the temperature indicators of the water are also considered unchanged.

According to this condition, Re $\approx$ P^{-1}—that is, the Re criterion is inversely proportional to the channel perimeter:

$$\frac{Re_k}{Re_0} \sim \frac{P_0}{P_k} = \frac{1}{2.93}.$$

$$Re_k = \frac{Re_0}{2.93} = \frac{19.45 \cdot 10^4}{2.93} = 6.64 \cdot 10^4 > 10^4.$$

When moving from a pipe to a rectangular channel, the flow regime of water does not change, therefore, the criterion equation remains unchanged.

We will study the effect of the equivalent diameter of the channel on the heat transfer intensity. From the criterion Eq. (11.2) it is known that: $\alpha \sim d_э^{-1} \cdot Re^{0.8} \sim P \cdot P^{-0.8} = P^{0.2}$, that is, when $F_0 = const$, as the perimeter surface of the channel increases, its equivalent diameter decreases, and the heat transfer intensity increases. Then, the heat transfer coefficient in the channel.

$$\frac{\alpha_k}{\alpha_0} \sim \left(\frac{P_k}{P_0}\right)^{0.2} = 2.93^{0.2} = 1.24$$

increases by 24%.

The heat flux change is calculated based on the above:

$$\frac{Q_k}{Q_0} \sim \frac{\alpha_k}{\alpha_0} \cdot \frac{F_k}{F_0} \sim \frac{\alpha_k}{\alpha_0} \cdot \frac{P_k}{P_0} \sim \left(\frac{P_k}{P_0}\right)^{1.2} = 2.93^{1.2} = 3.633.$$

From these derived solutions, it can be concluded that the heat transfer coefficient of the flow at the same cross-sectional area is 3.63 times greater for the heat transfer from a rectangular channel than for the heat transfer from a pipe. This is explained by the increase in the intensity of heat transfer in the channel and the increase in the perimeter of the cross-sectional area there.

If the flow of the heat carrier heated in the external environment is accompanied by a high-temperature heat transfer flow inside the pipe, and the flow is directed perpendicular to the pipe surface, the model for calculating the heat transfer process using standard equations is as follows.

11.3.1 Express the Calculation Results

The algorithm for setting and solving the problem of heat transfer during forced flushing of pipes and pipe bundles in the horizontal direction is carried out in a similar way to the general calculation method, with the form and composition of the heat transfer criterion Eq. (11.2) changing. The composition and form of these equations are selected from the reference literature for specific cases encountered in practice [15].

In our case, a boiling heat carrier flow moving in a straight pipe with an outer diameter $d = 32$ mm is cooled by a water flow in the horizontal direction (Fig. 11.2). In this case, if the water flow velocity $w_j = 4.0$ m/s, the average water temperature $t_j = 20$ °C, and the temperature on the pipe wall surface $t_w = 60$ °C, we determine the coefficient of heat transfer from the pipe wall to the cooling water.

To calculate the heat transfer of a cylindrical channel with a fluid flow transversely, the following criterion equations should be used [15]:

(1) For the case where $8 < Re_j < 1 \cdot 10^3$

$$Nu_j = 0.5 \cdot Re_j^{0.5} \cdot Pr_j^{0.38} \cdot \psi(Pr), \tag{11.16}$$

(2) For the case where $1 \cdot 10^3 < Re_j < 2 \cdot 10^5$

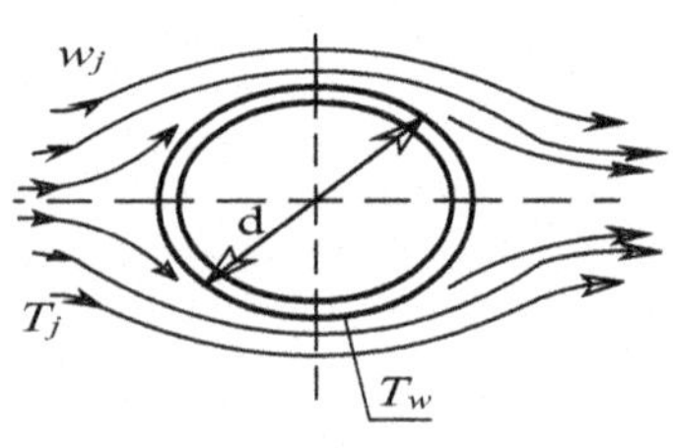

Fig. 11.2 Providing heat during transverse flushing of the pipe surface

$$Nu_j = 0.25 \cdot Re_j^{0.6} \cdot Pr_j^{0.38} \cdot \psi(Pr), \tag{11.17}$$

In Eq. (11.16), $\psi(Pr) = \left(\frac{Pr_j}{Pr_w}\right)^{0.25}$ is the correction for the change in the parameters of the heat carrier near the heat transfer surface; the diameter of the cylindrical body is taken as the defining parameter, the indices "j" and "w" denote the physical properties of the heat carrier and are selected based on the average temperature t_j of the moving flow and the average temperature t_w of the heat carrier on the heat exchange surface [17].

In this case, the flow regime when the pipe surface is washed by a water flow is determined. For the case where the water temperature $t_j = 20$ °C, the physical parameters of the heat carrier are selected from the reference [15]:

$$v = 1.003 \cdot 10^{-6} \frac{\text{m}^2}{\text{s}}; \quad \lambda = 59.85 \cdot 10^{-2} \frac{\text{W}}{(\text{m K})}; \quad Pr = 7.01.$$

In $t_w = 60\,°\text{C}$, $Pr_w = 2.99$.

The value of the number Re according to the selected parameters:

$$Re = \frac{w_j \cdot d}{v} = \frac{4 \cdot 0.032}{1.003 \cdot 10^{-6}} = 12.7 \cdot 10^4 > 10^4$$

indicates that the water flowing through the pipe is moving turbulently.

Now the heat transfer criterion equation can be solved. Because in the case under review

$$1 \cdot 10^3 < Re_j < 2 \cdot 10^5$$

and this indicates that expression (11.17) should be used.

Substituting exponents to solve the criterion equation gives the following result:

$$\begin{aligned} Nu_j &= 0.25 \cdot Re_j^{0.6} \cdot Pr_j^{0.38} \cdot \psi(Pr) \\ &= 0.25 \cdot \left(12.7 \cdot 10^4\right)^{0.6} \cdot 7.01^{0.38} \cdot \left(\frac{7.01}{2.99}\right)^{0.25} = 749.7. \end{aligned}$$

The value of the coefficient of heat transfer to water through the pipe wall:

$$\alpha = \frac{Nu_j \cdot \lambda}{d_1} = \frac{326.7 \cdot 0.5985}{0.032} = 14.021 \cdot 10^3 \frac{\text{W}}{(\text{m}^2\,\text{K})}.$$

11.4 Conclusion

Heat transfer is a physical process of heat (cold) transfer between the surface of solid bodies and the working environment (heat carriers) that wash them. Heat carriers consist of gases, liquids and solutions. Heat transfer is also a process that occurs as a result of convection and radiation heat exchange, as well as heat transfer.

The process of heat exchange through the surface area of heat exchange devices is the basis for the creation of all types of heat energy devices.

Regardless of whether the wall that separates the heat and the different temperature environments has a circular, flat, angular or flat shape, their materials will have different heat transfer properties. This parameter, in turn, depends on the temperature of heat carriers, wall thickness and a number of other thermal physical parameters.

In our case, the heat transfer process of heat carriers with high velocities at small temperature differences was modeled. According to the modeling results, it was found that the heat transfer coefficient depends mainly on the flow regime of the heat carriers, the Reynolds and Nusselt criteria, and the Prandtl number (Figs. 11.3, 11.4, 11.5, 11.6, 11.7 and 11.8).

Figure 11.3 shows a graph of the dependence of the heat transfer coefficient on the Nusselt criterion when the flow regime is laminar and turbulent, Fig. 11.4 the graph of the dependence of the heat transfer coefficient on the Nusselt criterion in laminar mode is depicted, Fig. 11.5 the graph of the dependence of the laminar flow regime on the Nusselt criterion is depicted when the heat carrier consumption is around 1 t/s, Fig. 11.6 the graph of the dependence of velocity on the Nusselt criterion in turbulent flow regime is depicted, Fig. 11.7 the graph of the dependence

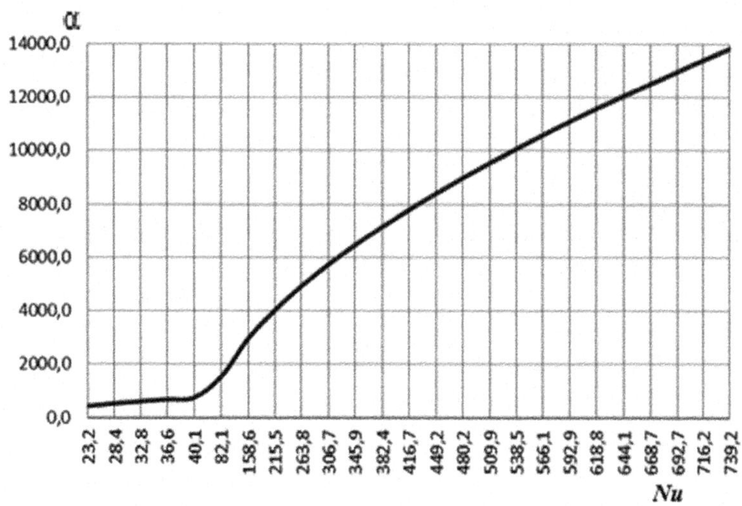

Fig. 11.3 The graph of the dependence of the heat transfer coefficient on the Nusselt criterion when the flow regime is laminar and turbulent

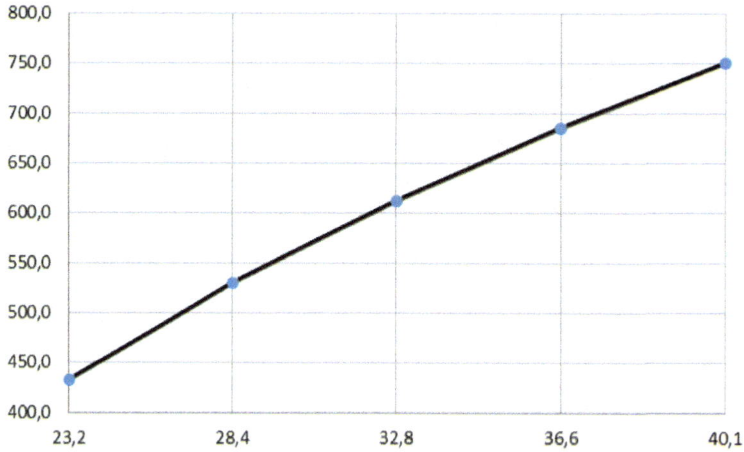

Fig. 11.4 Graph of the dependence of the heat transfer coefficient on the Nusselt criterion in laminar mode

Fig. 11.5 Graph of dependence of the laminar flow order on the Nusselt criterion when the heat carrier consumption is around 1 t/s

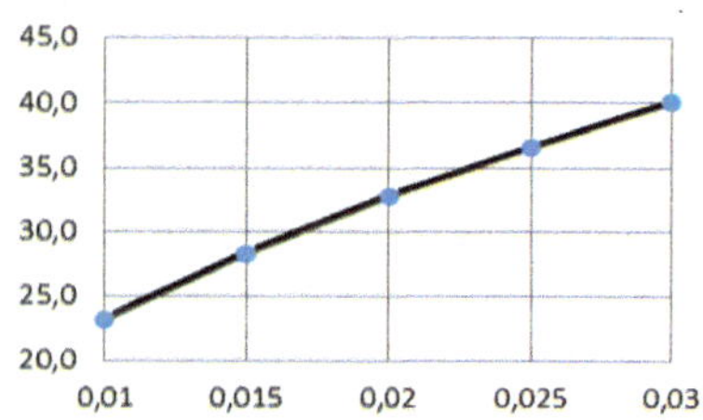

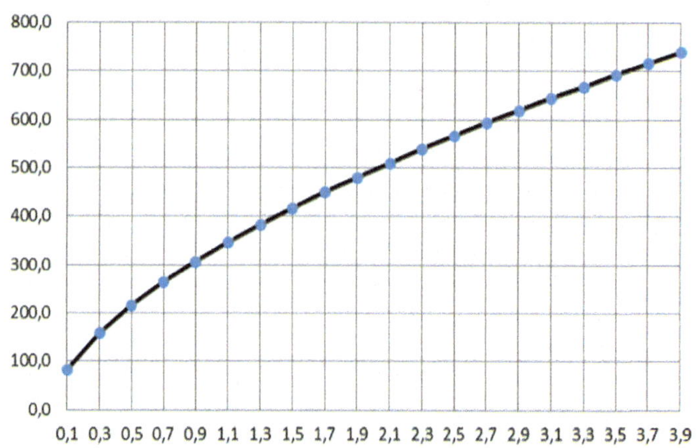

Fig. 11.6 The graph of the dependence of the speed on the Nusselt criterion in the turbulent flow regime

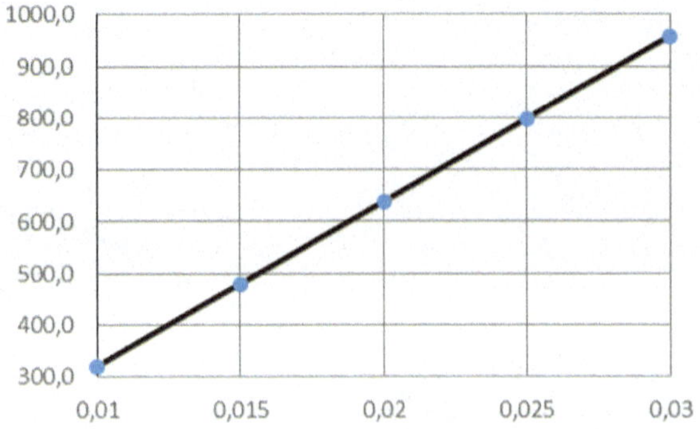

Fig. 11.7 Graph of the dependence of the velocity on the Reynolds number in the laminar flow regime

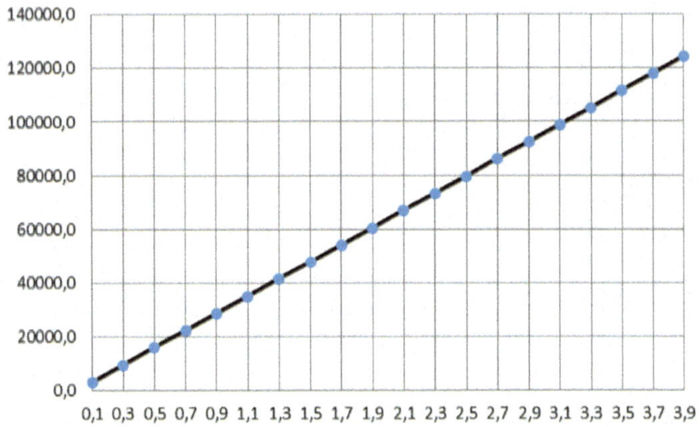

Fig. 11.8 Criterion (Reynolds) graph of the dependence of the speed of the heat carrier on the flow pattern

of velocity on Reynolds number in laminar flow regime is depicted and Fig. 11.8 Criterion (Reynolds) graph of the dependence of the speed of the heat carrier on the flow pattern is depicted.

The heat transfer coefficient of the heat carrier is used to describe the intensity of the heat exchange process. The numerical value of the heat transfer coefficient is equal to the heat flow generated per unit surface under the influence of temperature pressure (α, W/(m^2 K)), that is, it indicates the amount of heat transferred from the calculated surface unit in a unit of time. The heat transfer coefficient depends on the flow rate of the heat carrier, the type of flow, the geometry of the surface, etc.

References

1. S.M. Eldin, M. Alanazi, A. Alanazi, S. Alqahtani, S. Alshehery, A.E. Anqi, Economic and thermal analysis of a tubular thermoelectric power generator equipped with a novel fin-pin-porous based heat exchanger; comparative case study with conventional smooth channel. Therm. Eng. **48**, 103166 (2023)
2. T. Li, Y. Zhang, X. Gao, F. Jin, Combined heating and power generation performance of thermal station for electricity self-sufficient characterized by two-stage heat transfer. Therm. Eng. **53**, 103895 (2024)
3. G. Li, Q. Yao, Y. Hu, X. Fang, L. Wang, Investigating thermostat sensor offset impacts on operating performance and thermal comfort of three different HVAC systems in Wuhan. China. Therm. Eng. **31**, 101788 (2022)
4. M. Khaled, S. Ali, H. Jaber, J. Faraj, R. Murr, T. Lemenand, Heating/cooling fresh air using hot/cold exhaust air of heating, ventilating, and air conditioning systems. Energies **15**(5), 1877 (2022)
5. T. Mohapatra, B.N. Padhi, S.S. Sahoo, R.N. Pramanik, Performance analysis of three fluid heat exchanger used in solar flat plate collector system. Energy Procedia **109**, 322–330 (2017)
6. T. Mohapatra, S. Ray, S.S. Sahoo, B.N. Padhi, Numerical study on heat transfer and pressure drop characteristics of fluid flow in an inserted coiled tube type three fluid heat exchanger. Heat Transfer **48**(4), 1440–1465 (2019)
7. M.A. Sharaf, S. Marzouk, A. Aljabr, F.A. Almehmadi, T. Alam, D.A. Teklemariyem, Effects of multi-spring wires on hydrothermal performance of double tube heat exchanger. Therm. Eng. **60**, 104689 (2024)
8. A. Begag, R. Saim, S. Abboudi, H.F. Öztop, Effect of internal and external corrugated surfaces on the characteristics of heat transfer and pressure drop in a concentric tube heat exchanger. Int. J. Therm. Sci. **165**, 106930 (2021)
9. H.M. Maghrabie, K. Elsaid, E.T. Sayed, M.A. Abdelkareem, T. Wilberforce, M. Ramadan, A. Olabi, Intensification of heat exchanger performance utilizing nanofluids. Thermofluids **10**, 100071 (2021)
10. C. Ho, S. Huang, C.-M. Lai, Enhancing laminar forced convection heat transfer by using Al_2O_3/PCM nanofluids in a concentric double-tube duct. Therm. Eng. **35**, 102147 (2022)
11. G. Uzakov, S. Khamraev, Rural house heat supply system based on solar energy. IOP Conf. Ser.: Mater. Sci. Eng. **1030**(1), 012167 (2021)
12. S.S. Kutateladze, *Fundamentals of the Theory of Heat Transfer* (Atomizdat, 1979)
13. V.P. Isachenko, V.A. Osipova, A.S. Sukomel, *Heat Transfer* (Energoizdat, Moscow, 1981)
14. B. Toshmamatov, Development of automation and control system of waste gas production process based on information technology. Commun. Comput. Inf. Sci. **2112**, 196–205 (2024)
15. M.A. Mikheev, I.M. Mikheev, *Principles of Heat Transfer* (Energy, Moscow, 1973)
16. M.A. Mikheev, *Fundamentals of Heat Transfer* (Peace, Moscow, 1968)
17. H. John, V. Lienhard, *A Heat Transfer* (Phlogiston Press, Cambridge, MA, 2024)

Chapter 12
Mathematical Modeling of the Phenomenon of Electrical Conductivity of a Lithium-Ion Battery with a Large Capacity

Ibratjon Aliyev, Sultonali Abdurakhmonov, Ruzikhon Akhrorova, Moxizarxon Mamadaliyeva, and Nargiza Toirova

Abstract The paper presents a mathematical modeling of a storage unit based on a lithium-ion model. For the description, the problem is formulated on the basis of an inhomogeneous differential equation of electrical conductivity. In the course of the work, a pattern was derived that makes it possible to determine important aspects of the installation when entering boundary and initial conditions—parameters of the storage system. The result of the work was a proof of the possibility of improving modern lithium-ion batteries by combining component structures.

12.1 Introduction

The organization of new energy complexes designed for various, including large, capacities around the world creates a separate need for the organization of potential complexes capable of accumulating generated energy through new technologies. Today, storage devices are actively developing, along with the emerging demand. The mechanisms of electric energy accumulation were formed initially, using the properties of alkali metals and acids, to convert chemical energy into electrical energy,

I. Aliyev (✉)
SRI "PRNR", Electron Laboratory LLC, Margilan, Fergana Region, Republic of Uzbekistan
e-mail: alievibratzon12@gmail.com

S. Abdurakhmonov
Fergana Polytechnic Institute, Fergana, Fergana Region, Republic of Uzbekistan

R. Akhrorova
Andijan State University of Foreign Languages, Andijan, Fergana Region, Republic of Uzbekistan

M. Mamadaliyeva · N. Toirova
Fergana State University, Fergana, Fergana Region, Republic of Uzbekistan

© The Author(s), under exclusive license to Springer Nature Switzerland AG 2025
A. Gibadullin et al. (eds.), *Proceedings of the 4th International Scientific and Practical Symposium on Materials Science and Physics*, Springer Proceedings in Physics 318,
https://doi.org/10.1007/978-3-031-94805-3_12

which led to the use of galvanic cells. But a distinctive feature of storage systems was the possibility of additional battery charging [1].

And modern scientific capabilities allow us to use various models of such devices, with varying degrees of efficiency. Among the types of batteries, alkaline, metal-air, nickel–cadmium, calcium batteries, flow batteries, and others are distinguished. There are, of course, more exotic representatives of such devices from among gravitational and wave accumulating systems, but most of them are aimed at using chemical and physical processes directly. And among all currently known models, the lithium-ion battery can be fairly identified as the most promising.

This is due to the fact that modern technologies have not reached the limit of the possibility of such a design, due to the need to use various combinations in the design of a lithium-ion battery. The active metal refers to lithium as the element with the highest electrochemical activity.

And as practice shows, modern results on increasing the efficiency of batteries, which are about 20 Wh/kg among alkaline batteries and up to 30–40 Wh/kg among nickel–cadmium batteries, industrial lithium-ion batteries vary in their efficiency between 150 and 200 Wh/kg. Despite the fact that smaller or more advanced models used in electric vehicles or mobile electrical appliances can reach an efficiency value of about 250 Wh/kg. These data relate to liquid electrolyte batteries, which have some potential unacceptability to high temperatures. Current experiments and research show that the combination of additional elements in the separator, electrolyte, cathode, pure anode and collectors can lead to a greater increase in efficiency. An illustrative example of such an increase is the excess of 300 Wh/kg of solid-state electrolyte storage systems.

At the same time, the need for not only small but also industrial installations continues to grow due to the creation of new power plants, which follows from the growing demand of the population of countries, which in turn requires the provision of uninterrupted energy supply. For this reason, today the availability of a storage system for large power plants capable of retaining at least 50% of their capacity, especially for systems with their own electric grid, is becoming an important necessity, including in the territory of the Republic of Uzbekistan. Based on all of the above, as well as taking into account all these aspects, we can confidently conclude that the issue is relevant [2].

12.2 Materials and Methods of Research

In the course of the study, mathematical modeling methods were used by reducing electromagnetic interaction through modeling using a partial differential equation of thermal conductivity. Also, as an additional method, the possibilities of using functions as boundary and initial conditions are applied. The research materials were experimental, as well as statistical data of an empirical and theoretical nature.

12.3 Research

The use of a lithium-ion battery is based on methods of creating a potential difference between two parts of the terminals—between the anode and the cathode. The anode in this case is carbon graphite, which has a hexagonal crystal lattice in the form of a benzene tube, which is connected to a lithium ion. So, in this compound, lithium has 2 electrons in the first s-orbital and a single electron in the outer orbit. In the case of carbon atoms in the outer orbit, they have 2 electrons in the p-orbital, which allows them to form a full-fledged graphite compound and to embed lithium in it, one electron in the outer orbit becomes free, increases its degree of freedom, so that 2 electrons in the s-orbital become valence and the resulting compound is organized, intercalation is performed.

Cobalt dioxide acts as the cathode. In this case, cobalt with 27 electrons has the configuration—1s22s22p63s23p63d74s2, therefore it has 2 electrons in the outer orbit, while oxygen has 4 electrons in the outer orbit and 2 holes, due to a shortage in the outer p-orbital. This creates a potential difference in the local case of 2 elementary charges between lithium-carbon-6 and cobalt oxide [3].

Because of this, lithium is ionized, its electrons pass through the load and settle on cobalt dioxide. However, over time, the concentration of electrons begins to grow and the electrons acquire an obvious charge, where lithium-carbon lacks 1 electron, and cobalt dioxide now lacks 1 electron. Because of this, the created potential difference forces lithium ions that are not connected by a direct valence bond to the carbon crystal lattice to leave it and pass through the electrolyte to the side of the anode.

Because of this, a lithium-carbon-6 decay reaction is formed on the anode side into a benzene tube, lithium ion and electron (12.1), and on the cathode side, cobalt dioxide combines with an electron and lithium ion to form lithium-cobalt dioxide (12.2), where the reaction (12.3) is formed as a whole.

$$LiC_6 \rightarrow C_6 + Li^+ + e^- \tag{12.1}$$

$$CoO_2 + Li^+ + e^- \rightarrow LiCoO_2 \tag{12.2}$$

$$LiC_6 + CoO_2 \rightarrow C_6 + LiCoO_2 \tag{12.3}$$

The process continues from the point of view of free electrons until the concentration of electrons becomes large enough to stop them, but they are compensated by the remaining lithium ions, then the movement continues, but as a result, the potential difference does not become sufficient to get the last lithium ions. Thus, the lithium-ion battery is discharged. However, to prevent a short circuit and prevent the electrons and lithium ions from passing back during discharge, a separator is located in the electrolyte to stop the passage. Moreover, an alkali metal salt can be chosen as an electrolyte, and fiber, polymer film, ceramics or a natural compound of rubber, asbestos or wood can be used as a separator. But at the same time, it is important to note that in its pure form, when performing such a system, the passage of free electrons through the load will not be active enough–there will be a low current, therefore metal collectors are used: on the anode side–made of copper, on the cathode side–made of aluminum [4, 5].

Thus, the entire system consists of a series of plates: aluminum collector–cobalt dioxide cathode–alkali metal salt electrolyte–separator–alkali metal salt electrolyte–lithium carbon anode–copper collector. To carry out the reverse process, an external network is connected, from where charges flow to the anode, where they cause an increase in the number of free electrons in the carbon crystal lattice, which creates a stronger potential difference to attract lithium ions from the cathode. The ions return through the electrolyte, overcoming the separator due to the sufficient power of the potential difference to the anode, restoring the original appearance of the carbon compound. At the same time, this process is facilitated, since the potential difference created in the charger also draws electrons from the formed cobalt dioxide-lithium compound, which causes the compound to reverse intercalate for lithium ions. This is how the whole system works.

However, such a system has a number of disadvantages in its design, which lead to a loss of charge in the storage system over time. The reason lies in the reaction of lithium ions with the electrolyte, which leads to the formation of a solid interfacial electrolyte (TME). The second reason is the possibility of a reaction during the transition of lithium in excess to the anode with the formation of cobalt oxide and lithium oxide separately, which leads to a decrease in the amount of lithium and cobalt [6, 7].

As a result of this analysis, it becomes clear that the possibilities of such a lithium-ion battery design are quite wide and can be used in relation to various designs and modifications. But in order to simulate the process of changing components, using various types of electrolytes, separator elements, other metal oxides at the anode and collector parameters, it is necessary to simulate the existing system. For this, the parameter of the electrical conductivity of the entire system relative to lithium ions is initially used. This is done due to the possibility of establishing a relationship between electrical conductivity and a function of temperature, from a variable coordinate and time (12.4) [8].

$$\sigma = \sigma_0 e^{-\frac{\Delta W}{2kT}} \Rightarrow \begin{pmatrix} \Delta W = f(x,t) \\ \sigma = \sigma(x,t) \\ T = T(x,t) \end{pmatrix} \Rightarrow T(x,t) = -\frac{2k}{f(x,t)} \ln \frac{\sigma(x,\tau)}{\sigma_0} \tag{12.4}$$

If such a conversion is possible, the entire process should eventually be reduced back to the electrical conductivity parameter with all the consequences and reduced to the value of the efficiency of the storage system in Wh/kg. Thus, continuing the study in (12.4), it is possible to use the temperature function in an inhomogeneous partial differential equation of thermal conductivity with a known additional parameter (12.5).

$$\frac{\partial T(x,t)}{\partial t} = a^2 \Delta T(x,t) + f(x,t), a^2 = \frac{k(x,y,z)}{c(x,y,z)m(x,y,z)} \tag{12.5}$$

Boundary conditions must be set for the specified equation, however, in this case they can be set analytically due to the well-known design: collector–anode–electrolyte–separator–cathode–collector. So, the first boundary condition is a function along the coordinate (12.6), depending on the additional function of specific heat capacity and density function combined together (12.7).

$$T\left(x^*, t\right) = \frac{3W}{V}[c,\rho]\left(x^*\right) \tag{12.6}$$

$$[c,\rho](x) = \begin{cases} x = 0, 0 \\ 0 < x \leq l_1, (c_1\rho_1)^{-1} \\ l_1 < x \leq l_2, (c_2\rho_2)^{-1} \\ l_2 < x \leq l_3, (c_3\rho_3)^{-1} \end{cases} \tag{12.7}$$

Deriving the initial condition for a given problem with a known domain of definition is a more difficult task. Because if a single moment of time is taken in the case of a boundary condition, it will mainly depend on the power parameter. Moreover, in the case of a coordinate, the temperature is determined based on the ratio of heat capacity and 3 acts as the number of elements in this consideration. As the number of layers increases, this value changes, which makes it possible to calculate the arithmetic mean coefficient when switching from the energy value to the temperature value.

In the case of the initial boundary condition, it is necessary to trace the entire path of lithium ions, which change the temperature and electrical conductivity of the entire system and all its components. However, in this case, the voltage is known—the electrochemical activity of lithium ions, and the resistance is also known as the ratio of the lengths of all plates per area, along with the arithmetic mean of the electrical resistivity. The next parameter is the number of lithium-ion charges, which can be determined based on their total current strength, and then from the definition of current strength as the product of the number of charges per elementary charge, the cross-sectional area of the ion path and their average velocity (12.8).

$$U = E_{Li}, R = \frac{l_3^2}{3V}\sum_{i=1}^{3}(\rho_e)_i, n = \frac{U}{Re\frac{V}{l_3}\sqrt{1-\left(\frac{1}{1+\frac{eE_{Li}}{m_e c^2}}\right)^2}} \tag{12.8}$$

Thus, based on all of the above, a separate power value of lithium ions (12.6) is created, on which the temperature indicator depends over time and on which, taking into account that this energy reaches its peak starting from zero and then decreases, one can see an analogy with oscillatory motion of a sinusoidal nature, where as an amplitude a certain power appears, thus forming the analytical form of the initial condition of the problem (12.9).

$$W_{Li^{+}} = 3E_{Li}^{2} l_{3}^{-1} \sum_{i=1}^{3} (\rho_{e})_{i}^{-1} \left(1 - \left(\frac{1}{1 + \frac{eE_{Li}}{m_{e}c^{2}}}\right)^{2}\right)^{-\frac{1}{2}} \tag{12.9}$$

$$T\left(x, t^{*}\right) = \frac{3W_{Li^{+}}}{V} \sin\left(\frac{W_{Li^{+}}R}{U^{2}} t^{*}\right) \tag{12.10}$$

Thus, with a known domain of definition, Eq. (12.5) with the desired function of coordinate and time, boundary (12.6) and initial condition (12.10), the problem can be solved using the Fourier variable separation method. To do this, the final form of Eq. (12.11) is initially indicated and the temperature function is presented as a product of a separate function from the coordinate and a separate function from time (12.12).

$$\frac{\partial T(x, t)}{\partial t} = a^{2} \frac{\partial T(x, t)}{\partial x} + W \tag{12.11}$$

$$T(x, t) = X(x)Y(t) \tag{12.12}$$

The next stage is to substitute a new form of the function into the equation, where to separate the variables, you have to use the logarithm of both sides of the identity. After that, through the properties of the logarithm, the components are separated, which also leads to a reduction in the terms of the formed polynomial, forming a finite expression with only the time function and its derivative on the left side of equality and the same analogy on the right side of equality. This identity is equated to the first independent variable of the problem, which makes it possible to form two additional ordinary partial differential equations with respect to each of the functions (12.13).

$$\begin{aligned} X(x)Y'(t) = a^{2}X'(x)Y(t) + W \Rightarrow \frac{Y'(t)}{Y(t)} \\ = a^{2}\frac{X'(x)}{X(x)} + \frac{W}{X(x)Y(t)} \\ \Rightarrow \ln\frac{Y'(t)}{Y(t)} = \ln a^{2}\frac{X'(x)}{X(x)} + \ln\frac{W}{X(x)Y(t)} \\ = \ln a^{2}\frac{X'(x)}{X(x)} + \ln W - \ln X(x) - \ln Y(t) \\ \Rightarrow \ln\frac{Y'(t)}{Y(t)} + \ln Y(t) = \ln a^{2}\frac{X'(x)}{X(x)} + \ln W - \ln X(x) = -\lambda \end{aligned} \tag{12.13}$$

The solution of the first equation occurs by means of the simplest integration to obtain the general form (12.14), and the substitution of the initial conditions in analytical form for the first case with zero forms the expression for the first independent variable, which eventually forms the final form of the time function (12.15).

$$\ln \frac{Y'(t)}{Y(t)} + \ln Y(t) = -\lambda \Rightarrow \int dY(t) = e^{-\lambda} \int dt \Rightarrow Y(t) = te^{-\lambda} + C \tag{12.14}$$

$$\begin{cases} 0 * e^{-\lambda} + C = \frac{3W_{Li+}}{V} \sin\left(\frac{W_{Li+}R}{U^2} * 0\right) => C = 0 \\ \lambda = -\ln\left[\frac{3W_{Li+}}{Vt} \sin\left(\frac{W_{Li+}R}{U^2} t^*\right)\right] \end{cases}$$

$$\Rightarrow Y(t) = te^{-\ln\left[\frac{3W_{Li+}}{Vt} \sin\left(\frac{W_{Li+}R}{U^2} t^*\right)\right]} \tag{12.15}$$

The second ordinary differential equation of the first degree, depending on the coordinate function, can already be equated to the resulting form of the first independent variable of the problem, which will allow dividing the equation to the state of direct integration to obtain the general form (12.16).

$$\begin{aligned} &\ln a^2 \frac{X'(x)}{X(x)} + \ln W - \ln X(x) = \ln\left[\frac{3W_{Li+}}{Vt} \sin\left(\frac{W_{Li+}R}{U^2} t^*\right)\right] \\ &\Rightarrow \ln a^2 + \ln X'(x) + \ln W = \ln\left[\frac{3W_{Li+}}{Vt} \sin\left(\frac{W_{Li+}R}{U^2} t^*\right)\right] \\ &\Rightarrow a^2 X'(x) W = \frac{3W_{Li+}}{Vt} \sin\left(\frac{W_{Li+}R}{U^2} t^*\right) \\ &\Rightarrow \int dX(x) = \int \frac{3W_{Li+}}{a^2 WVt} \sin\left(\frac{W_{Li+}R}{U^2} t^*\right) dx \\ &\Rightarrow X(x) = \frac{3W_{Li+}x}{a^2 WVt} \sin\left(\frac{W_{Li+}R}{U^2} t^*\right) + C \end{aligned} \tag{12.16}$$

Due to the zero-component relative to the specific heat and density function in the analytical notation of the boundary condition, the simplest derivation of the function from coordinate (12.17) is obtained.

$$\begin{aligned} &\frac{3W_{Li+} * 0}{a^2 WVt} \sin\left(\frac{W_{Li+}R}{U^2} t^*\right) + C = \frac{3W}{V} [c, \rho](0) \Rightarrow C = 0 \\ &\Rightarrow X(x) = \frac{3W_{Li+}x}{a^2 WVt} \sin\left(\frac{W_{Li+}R}{U^2} t^*\right) \end{aligned} \tag{12.17}$$

Thus, based on the shape of the time function (12.15) and the coordinate function (12.17), the resulting temperature function (12.18) is formed, which, by substitution, is transformed into the form of the function for electrical conductivity (12.19).

$$T(x,t) = \frac{3W_{Li^+}x}{a^2 WVt} \sin\left(\frac{W_{Li^+}R}{U^2} t^*\right) t e^{-\ln\left[\frac{3W_{Li^+}}{Vt} \sin\left(\frac{W_{Li^+}R}{U^2} t^*\right)\right]} \tag{12.18}$$

$$\sigma = \sigma_0 \exp\begin{pmatrix} -\frac{\Delta W}{2k} * \frac{a^2 WVt}{nW_{Li^+}xt} \sin^{-1}\left(\frac{W_{Li^+}R}{U^2} t^*\right) * \\ * \exp\left(\ln\left[\frac{3W_{Li^+}}{Vt} \sin\left(\frac{W_{Li^+}R}{U^2} t^*\right)\right]\right) \end{pmatrix} \tag{12.19}$$

As a result of the research, a time function and an electrical conductivity function were obtained that can fully simulate storage systems.

12.4 Results

To form conclusions about the practical results obtained, it is necessary to carry out some modeling and representation in the form of graphs of the obtained equations. Thus, we will install a system consisting of an aluminum collector on an anode with a thickness of 1.5 mm, an anode made of lithium after intercalation in carbon with a thickness of 2.5 mm, an electrolyte made of dilute 10% sulfuric acid with a layer thickness of 2 mm, a separator made of natural organic fiber with a thickness of 1 mm, a cobalt oxide cathode with a thickness of 2.5 mm and a copper collector on The cathode is 1.5 mm thick. At the same time, the width of the system is 50 cm and the height is 60 cm with a total volume of 3300 cm^3. A charging power of about 500 kWh is sent to this model, which creates a flow of lithium ions with a voltage of –3.04 V and a resistance of about 10 V.

Thus, such a system will have different temperatures in different areas, which is reflected in the corresponding three-dimensional graph (Fig. 12.1), as well as in the graph of electrical conductivity (Fig. 12.2).

From the graph obtained, it is clearly seen that in this case the entire installation is divided by a temperature gradient into an explicit number of parts, which have their own special forms of change, similar to an altered parabola or a logarithmic graph in a time projection. The last property of the function can be seen from the second projection of the three-dimensional graph (Fig. 12.3).

The overall temperature is between 100 and 215 °C. Regarding the level of electrical conductivity, according to the percentage representation of the graph, electrical conductivity acts at almost the maximum possible level due to the possibility of conducting the specified large amount of energy during input. Special attention should be paid to the formulation of the model at different levels of incoming energy, with the indicated dimensions of the structures of the storage system.

So, with a decrease in energy, there is a decrease in temperature, but the difference in temperature heights begins to change, as well as the degree of approximation to the logarithmic model when projecting the lateral part. This is clearly seen when the energy is reduced to a state of 2 J, relative to temperature (Fig. 12.4).

As the power decreases, the electrical conductivity becomes almost uniform, becoming a constant state, and the only distinguishing feature is the obvious peaks in

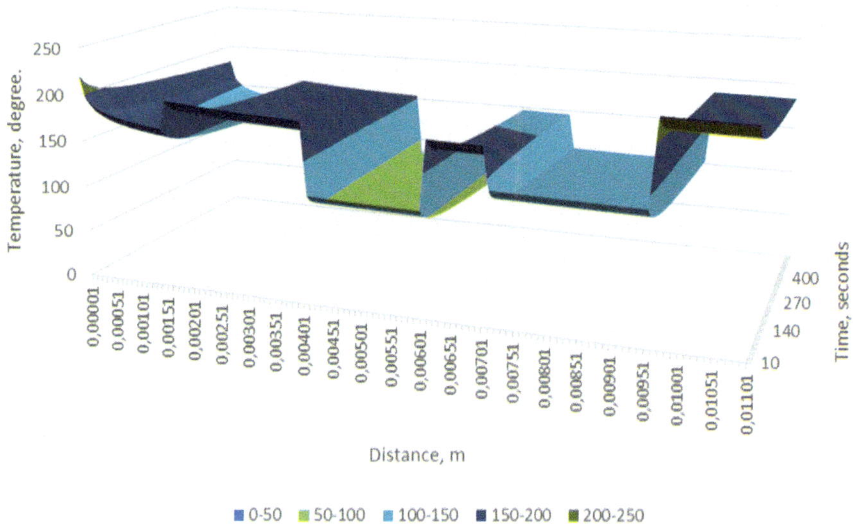

Fig. 12.1 Three-dimensional graph of temperature rise of the standard model

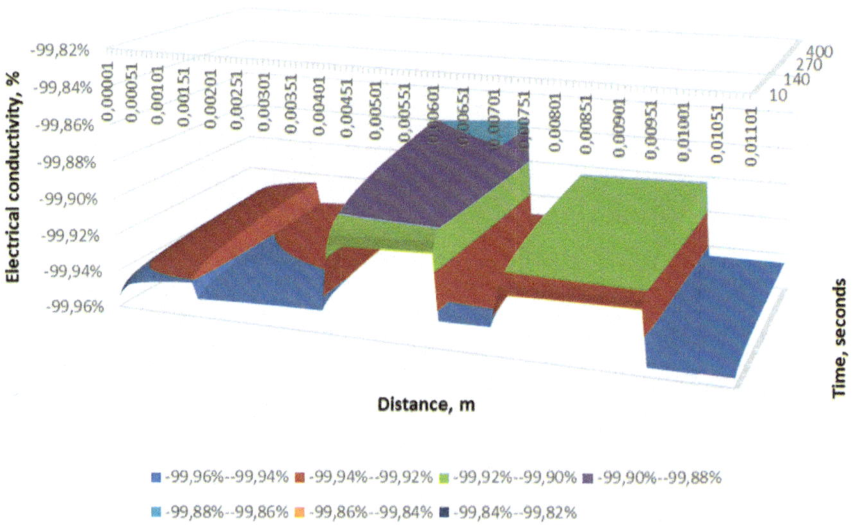

Fig. 12.2 Three-dimensional graph of electrical conductivity as a percentage

some areas of the graph. Such sharp increases, visible on a small scale in the range of 300–400 s at the cathode and copper collector levels, in the middle, at the separator level as a single peak and almost imperceptible peaks in the anode region, can be offset by quantum fluctuations involved in the process (Fig. 12.5).

However, one of the goals initially set was a practical test of the possibility of using the structure at various dimensions for large capacities. So, this can be most

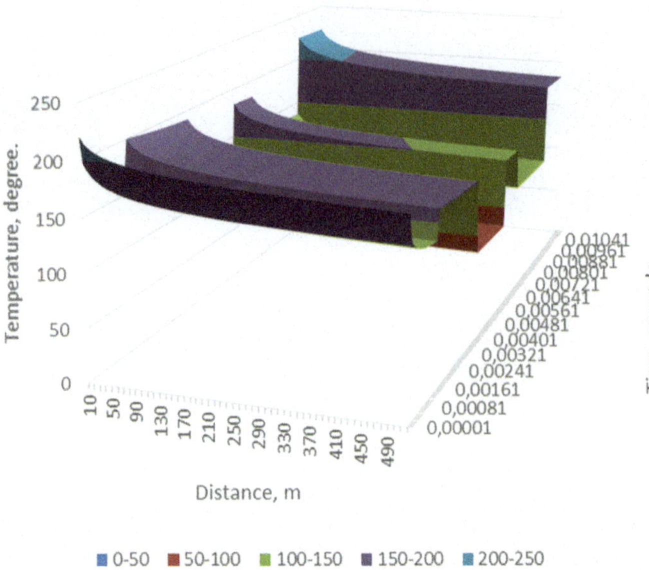

Fig. 12.3 The second projection of the temperature graph from the coordinate and time

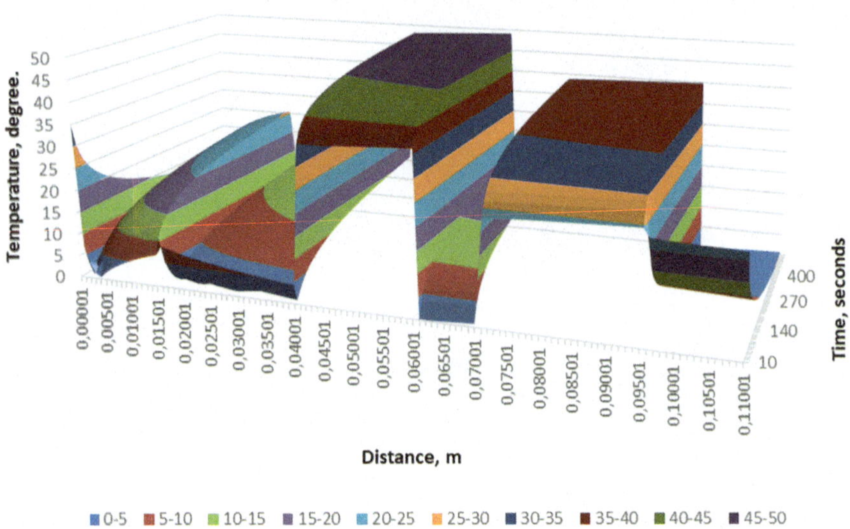

Fig. 12.4 Three-dimensional graph of temperature at low power

clearly seen with an increase in power on the graph of electrical conductivity, where it has reached 0.16% in a percentage ratio that used to be in the range of 100%, tending to 0%. This result is achieved for an installation measuring 25 m wide and 50 m long, with an energy output of 2.4 GWh. The result obtained on the graph means that

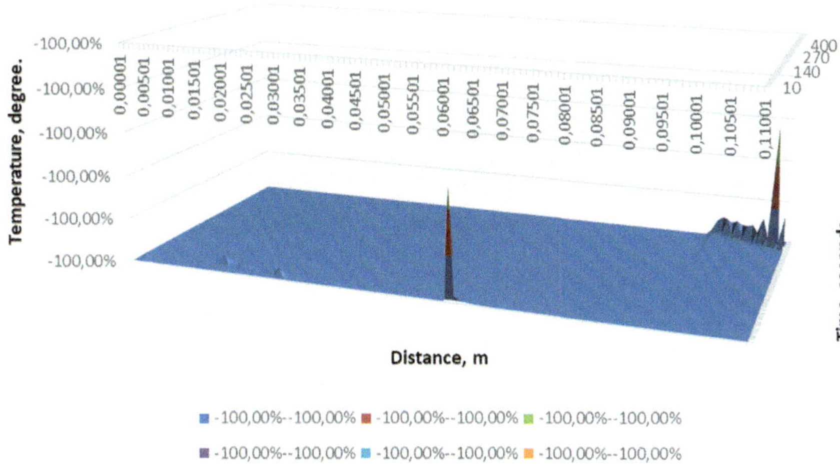

Fig. 12.5 Three-dimensional graph of electrical conductivity

in this case the electrical conductivity increases critically, which leads to a corona discharge in the separator and anode, where the maximum electrical conductivity is observed on the corresponding graph (Fig. 12.6).

In terms of temperature, there are no sharp peaks in this case, although the function becomes reversed, and the entire shape of the graph begins to grow to 250° (Fig. 12.7).

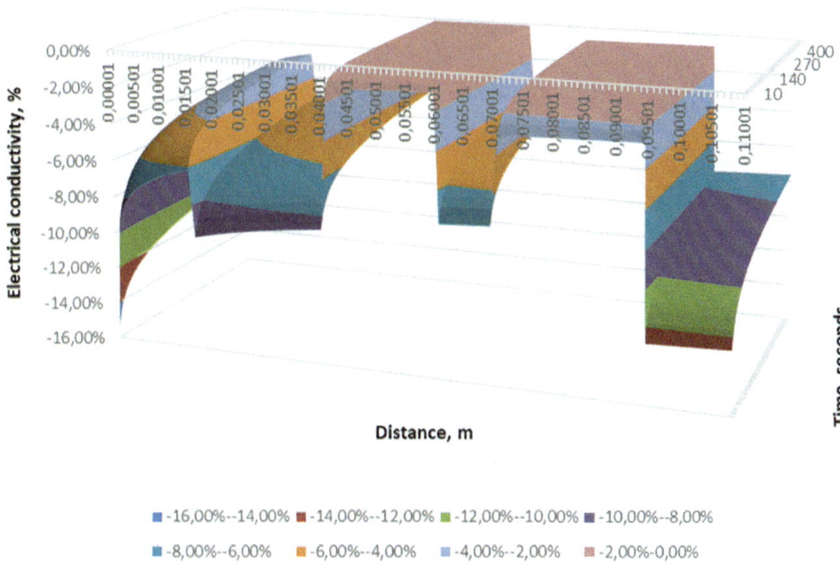

Fig. 12.6 Electrical conductivity graph at maximum power

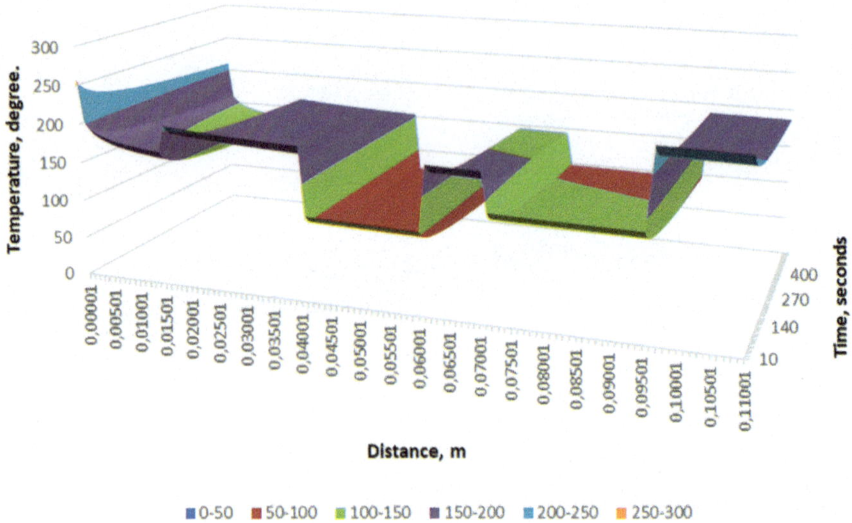

Fig. 12.7 Graph of the temperature of the storage system at peak power

Thus, the formed model, with a known size, design, chemical components and physical properties, can reach a critical value of the order of 2.4 GWh, while having an overall efficiency of 5760.88 Wh/kg in a critical case. For comparison, the current performance of standard battery systems is 150–200 Wh/kg for industrial models and 200–250 Wh/kg for standard design types. This proves that an increase in the purity of the chemical composition, as well as the possibility of combining the types of various components of the storage system, including using a mathematical model, can lead to significantly greater efficiency results.

12.5 Conclusion

In the course of the research, a full-fledged mathematical model was developed that can dynamically take into account changes in the entire design of the collector–anode–electrolyte–separator–cathode-collector storage system. At the same time, during the study, the installation was modeled in the standard case, with low power and in critical form, where the efficiency results of 5760.88 Wh/kg were clearly demonstrated, which is 28.8 times higher than the current results.

The achievement of the work was the proof of the need to use various compounds as a cathode, separator, anode and collectors, leading to the greatest results with greater efficiency. Thus, a highly focused mathematical apparatus was created that is able to fully predict the results of the efficiency of private and industrial scale storage facilities.

References

1. P. Siano, Demand response and smart grids—a survey. Renew. Sustain. Energy Rev. **30**, 461–478 (2014). https://doi.org/10.1016/j.rser.2013.10.022
2. B. Kang, G. Ceder, Battery materials for ultrafast charging and discharging. Nature **458**(7235), 190–193 (2009). https://doi.org/10.1038/nature07853
3. A. Mauger, C.M. Julien, Critical review on lithium-ion batteries: are they safe? Sustainable? Ionics **23**(8), 1933–1947 (2017). https://doi.org/10.1007/s11581-017-2177-8
4. R. Zhang, S. Fujimori, The role of transport electrification in global climate change mitigation scenarios. Environ. Res. Lett. **15**(3), 034019 (2020). https://doi.org/10.1088/1748-9326/ab6658
5. N.A. Godshall, I.D. Raistrick, R.A. Huggins, Thermodynamic investigations of ternary lithium-transition metal-oxygen cathode materials. Mater. Res. Bull. **15**(5), 561 (1980)
6. J.O. Besenhard, H.P. Fritz, Cathodic reduction of graphite in organic solutions of alkali and NR4+ salts. J. Electroanal. Chem. Interfacial Electrochem. **53**(2), 329–333 (1974)
7. M.M. Thackeray, J.O. Thomas, M.S. Whittingham, Science and applications of mixed conductors for lithium batteries. MRS Bull. **25**(3), 39–46 (2011). https://doi.org/10.1557/mrs2000.17
8. X. Kang, Nonaqueous liquid electrolytes for lithium-based rechargeable batteries. Chem. Rev. **104**(10), 4303–4418 (2004). https://doi.org/10.1021/cr030203g

The manufacturer's authorised representative in the EU is Springer Nature Customer Service Centre GmbH, Europaplatz 3, 69115 Heidelberg, Germany. If you have any concerns regarding our products, please contact ProductSafety@springernature.com

Printed and bound by CPI Group (UK) Ltd, Croydon, CR0 4YY

30/09/2026

02225471-0018